Advanced Euclidean Geometry

Roger A. Johnson

Dover Publications, Inc.
Mineola, New York

Bibliographical Note

This Dover edition, first published in 1960 and republished in 2007, is an unabridged republication of the first edition, originally published under the editorship of John Wesley Young by Houghton Mifflin Company, Boston, in 1929 under the title *Modern Geometry*. The 1960 Dover edition was published under the title and subtitle *Advanced Euclidean Geometry: An Elementary Treatise on the Geometry of the Triangle and the Circle.*

Library of Congress Cataloging-in-Publication Data

Johnson, Roger A. (Roger Arthur)
   [Modern geometry]
   Advanced Euclidean geometry / Roger A. Johnson.
      p. cm.
   "This Dover edition, first published in 1960 and republished in 2007, is an unabridged republication of the first edition, originally published under the editorship of John Wesley Young by Houghton Mifflin Company, Boston, in 1929 under the title Modern Geometry. The 1960 Dover edition was published under the title and subtitle Advanced Euclidean Geometry: An Elementary Treatise on the Geometry of the Triangle and the Circle."
   Includes index.
   ISBN-13: 978-0-486-46237-0
   ISBN-10: 0-486-46237-4
   1. Geometry, Modern—Plane. 2. Geometry, Plane. I. Title.

QA474.J6 2007
516.2—dc22

                                            2007011655

www.doverpublications.com

EDITOR'S INTRODUCTION

THERE are fashions in mathematics as well as in clothes — and in both domains they have a tendency to repeat themselves. During the second half of the nineteenth century "Modern Geometry," in the sense of the content of the present book, aroused much interest and was prosecuted vigorously by a considerable number both in England and on the continent of Europe. Many beautiful new theorems were proved, most of them by elementary methods. Toward the end of the century this interest waned somewhat.

The present appears to exhibit a revival of this interest. This is in large part due to the recognition of the value of this new material as training for the prospective teacher of geometry in our secondary schools. Here is indeed a discipline which is a natural "sequel" to elementary geometry, a body of propositions which may be derived by methods similar to those used in the classical plane geometry and which has all the attraction of novelty and inherent beauty. It is not surprising, therefore, that an increasing number of colleges and normal schools are introducing into their curricula courses in this "Modern Geometry."

It is not only as a textbook in such courses, however, that the present book will be a valuable addition to our mathematical literature. In view of the very modest demands it makes on the previous training of the student, it may be expected to appeal to many teachers in secondary schools and colleges whose interest and ambition lead them to seek to increase their knowledge and appreciation of geometry. Moreover, many highly trained mathematicians will welcome it as giving them the opportunity to fill a not uncommon gap in their previous study. The content of this book, in spite of its elementary character, is by no means well known to mathematicians in general.

Finally, the author has succeeded in incorporating a great deal of material which is scattered through journals or is otherwise not easily accessible, and this without impairing its value as an elementary text. It should, therefore, be found valuable also as a reference work.

J. W. YOUNG

PREFACE

This book deals with the geometry of the triangle and the circle, as developed extensively in the nineteenth century by British and Continental writers. This geometry, based entirely on the elementary plane geometry of Euclid or its modern equivalent, is rapidly coming to its due recognition as excellent material for college courses. Perhaps in no other field is there so rich a harvest of geometric truth so directly accessible to the student, with as little preliminary investment of energy in the development of method and technique. The student who is familiar with high school mathematics and with the language of trigonometry is well qualified to reap the full benefit of a course in this subject. For this reason, such a course makes a strong appeal to the teacher or the prospective teacher of mathematics in the secondary school; to the general student who appreciates mathematics, especially geometry, but is not attracted by the arduous algebraic difficulties of analytic geometry; and to the mathematician who finds frequent applications of this modern elementary geometry in other fields of mathematics. Such relationships are occasionally suggested or implied in the book; so that the reader who is acquainted with higher geometry will frequently recognize familiar theorems more or less effectually disguised.

The study of the recent elementary geometry may be carried on in various ways. Some of the writers have used freely the projective methods of central projection and of anharmonic ratios; and another method of attack is the analytic, making use of trilinear coördinates. The point of view of this book is that the subject matter deals exclusively with the elementary concepts associated with congruent and similar figures; that the synthetic projective or the analytic method of treatment implies more elaborate basic concepts, invariant with regard to the projective group of transformations; and that it is more elegant and far more appropriate to use only the Euclidean relations of congruence and similarity. Thus a directness and unity of treatment is achieved, that seems to be lost when the more powerful methods of higher geometry are introduced. In the following pages, then, we shall confine ourselves to the study of equal and similar figures, both in the formulations and in the proofs of theorems.

It may be thought that the liberal use which we shall make of circular inversion is a violation of this unity; but while it is true that the geometer may regard inversion as a quadratic Cremona transformation, it is equally easy and natural to define it in terms of similar figures and proportions, thus justifying its introduction and use.

The material which appears in this book is for the most part to be found in standard sources, many of them easily accessible. The most important of these are the following:

SIMON, MAX: *Ueber die Entwickelung der Elementargeometrie im XIX Jahrhundert.* Berlin, 1906.

> (A summary, with very full bibliography, of the most important modern developments of geometry; very useful for reference.)

CASEY, JOHN: *A Sequel to Euclid.* Dublin, 1881, 1888.

> (Four editions of this famous work, the first issued in 1881, were followed in 1888 by a fifth, which contained an 80-page "supplementary chapter" treating of some of the Brocard geometry. After Casey's death in 1891, a sixth edition, entitled "Part I" and without the supplementary chapter, was issued. The fifth, therefore, is the most interesting edition of the book; but since the material is all available elsewhere, the interest in the book is mainly historical. The author is indebted to Dr. R. C. Archibald for the loan of a copy of the comparatively rare fifth edition.)

LACHLAN, R.: *Modern Pure Geometry*. London, 1893.

MCCLELLAND, W. J.: *Geometry of the Circle*. London, 1891.

RUSSELL, J. W.: *Elementary Pure Geometry*. Oxford, 1893.

DURELL, C. V.: *Modern Geometry*. London, 1920.

GALLATLY, W.: *Modern Geometry of the Triangle*. London, 1910.

(These books bear some resemblance to one another, dealing with various aspects of geometry, usually by projective methods.)

COOLIDGE, J. L.: *A Treatise on the Geometry of the Circle and the Sphere*. Cambridge (England), 1914.

(Chapter 1 of this book is a survey of our field. The following chapters treat the geometry of the circle and sphere analytically and very fully, with many illuminating associations with the elementary field.)

FUHRMANN, W.: *Synthetische Beweise Planimetrischer Sätze*. Berlin, 1890.

EMMERICH, A.: *Die Brocard'schen Gebilde*. Berlin, 1891.

(Two valuable German treatises. The second deals with the Brocard geometry, in part by analytic methods; the first is freely quoted in the present text.)

ALTSHILLER-COURT, N.: *College Geometry*. Richmond, 1923.

(A recent and successful American text with which we expect to enter into friendly competition.)

An attempt has also been made to explore the very large number of journal articles, as well as less familiar books, and to incorporate in the text the most important results to be found in these sources. Since our book is not intended as an exhaustive treatise, but as an introduction to this very extensive field, the hundreds of more elaborate researches to be found in the periodicals find here little place.[1] While the

[1] Among the most important contributors to the geometry of the triangle must be named Dr. John S. Mackay, of Edinburgh, the first President of the Edinburgh Mathematical Society. Dr. Mackay was an enthusiastic worker in this field, and during the first twenty years of the existence of the Society he published in its *Proceedings* some thirty-five articles varying from brief notes to long monographs on the most important configurations related to the triangle. His historical researches also were of the greatest value, as will

subject matter of this book lays little claim to originality, and the original contributions of the author are of no great importance, he has devoted himself to the weaving of the material into an organic unity and to the strengthening, simplification, and clarification of the proofs. If the reader feels an æsthetic satisfaction in the unity and harmony of the arrangement and the interrelation of the various parts to one another, the author will have been successful.

Perhaps the chief contribution of the book to the advancement of the art of geometry is the concept and the method of proof designated, for lack of a better name, as that of "directed angles." The advantages of this method, already indicated in articles in *The American Mathematical Monthly* some years ago, can be appreciated only by full acquaintance with it, and one may hope that it will find its way into general use. Besides its power as a method of proof, it furnishes a valuable form of statement for certain fundamental theorems, which would otherwise require several statements for different cases. Such characteristic theorems as **75, 186,** and **238** have hitherto been haltingly and equivocally formulated in some of the texts; but their full significance stands out clearly when expressed in this symbolism. This new and rigorous method is submitted for the consideration of all geometers.

Undoubtedly the book offers more material than can be treated in the usual semester course. In the dilemma between omission of material and brevity of proof, the author has leaned toward the latter alternative; thus comparatively few theorems are proved in complete detail, and the amount of original work left for the student to complete is indeed extensive. At the same time, it is believed that the logical sequence is clear throughout, so that the reader will seldom be

be seen later. The student who has completed the present work and desires to pursue further studies in the same field can hardly make better progress than by reading the articles of Mackay; in these, moreover, he will find full bibliographical references.

puzzled by any real difficulty. It is expected that the reader will supply proofs of all theorems and corollaries not proved in the text; and hints are supplied where needed. Careful drawings are also of the greatest importance; and it is expected that the student will draw figures to illustrate the more important theorems.[1]

It is believed that teachers will find it possible to select material for a course of any length in accordance with their personal tastes, without impairing the unity of the work as a whole. The pivotal chapters, essential to any study of the subject, are: I, II, III, VII–XI, and the indicated portions of IV, V, and XII. Neither the geometry of circles (V, VI) nor the Brocard geometry (XII, XVI, XVII, XVIII) should be slighted; and Chapter XIV while not indispensable, gives a valuable insight into the essence of earlier chapters.

The author takes this opportunity to express his debt to Professor J. L. Coolidge, of Harvard University, in whose course in the Geometry of the Circle he was first introduced to this domain; and whose kindly and genial interest has been sustained during the preparation of this work. At the same time it should be made clear that Professor Coolidge is to be exonerated of any direct culpability for this production.

The thanks of the author are also due to Professors J. W. Young (the editor of the series) and B. H. Brown, both of Dartmouth College, for their painstaking reading of the manuscript and for many valuable suggestions; and to Professor R. C. Archibald, of Brown University, for a number of equally useful suggestions.

[1] See **14**, page 10.

CONTENTS

CONTENTS

MODERN GEOMETRY

CHAPTER I

INTRODUCTION

1. Prerequisites. It is assumed that the reader is acquainted with the elementary algebra and plane geometry ordinarily taught in American high schools, and with the simplest principles of trigonometry. Some familiarity with the standard theorems of plane geometry is assumed, and the reader may do well to review this subject before undertaking the work before us. Simple algebraic methods of reduction and manipulation will frequently be used; and the expression of geometric relationships will frequently be simplified by introducing the trigonometric functions, and occasionally by utilizing the most elementary identities connecting them. Indeed, some high-school courses in mathematics introduce as much trigonometry as is essential to our needs, and the study of geometry is greatly facilitated by the free use of algebraic and trigonometric methods. No further acquaintance with mathematics than this will be assumed, though the reader who is acquainted with higher geometry will frequently perceive the bearing of this work on other fields of geometry.

In the present chapter we set forth a number of general principles, methods, and points of view which will be adopted throughout the work. The more advanced student of mathematics will find little that is novel in these principles, while the reader to whom they are new will encounter no serious difficulty with them.

2. When we have occasion to deal with a geometric quantity that may be regarded as measurable in either of two directions, it is often convenient to regard measurements in one of these directions as positive, the other as negative. A familiar example is the thermometer scale.   Again, in measuring distances along an east-west road, we may attach to all eastward distances the positive sign, and to those westward the negative sign.   Then the resulting distance and direction from the starting point, achieved by combining two or more journeys on the road, whether in the same direction or not, is found by taking the algebraic sum of the numbers that represent the journeys.   Similar examples will suggest themselves; the essential principle is that the combination of the quantities is represented by the algebraic addition of their measures, as defined in the following paragraph.

The most important instance of this principle is that of distances along a straight line.   *If A and B are any two points, $\overline{AB}$ means the distance from A to B, and $\overline{BA}$ the distance from B to A.*   One of these will be represented by a positive number, the other by the same number with the negative sign.   For any three points A, B, C on a line, we have then the following important relations:

$$\overline{AB} + \overline{BA} = 0$$
$$\overline{AB} + \overline{BC} = \overline{AC}$$
$$\overline{AB} + \overline{BC} + \overline{CA} = 0$$
$$\overline{BC} = \overline{AC} - \overline{AB}$$

It may be noted that the last three are alternative versions of the same fact.   Special attention is to be given to the last, which expresses the directed distance between two points in terms of their distances from a fixed point of reference A. This suggests a useful method, whereby distances among the points of a line may be thus expressed, and relations among them may be established by means of algebraic relations. If the distances from a point O of the line $ABC\ldots$ are given by

$$a = \overline{OA}, \, b = \overline{OB}, \text{ etc.,}$$

then AB is represented by $(b - a)$, and so on.   For example:

3. Euler's Theorem.   *For any four points A, B, C, D of a line,*

$$\overline{AB}\cdot\overline{CD} + \overline{AC}\cdot\overline{DB} + \overline{AD}\cdot\overline{BC} = 0$$

For $(b - a)\,(d - c) + (c - a)\,(b - d) + (d - a)\,(c - b)$
$$\equiv bd - ad - bc + ac + \text{etc.} \equiv 0$$

4. With regard to angles, it is the accepted convention that *angles measured in the direction opposite to the motion of the hands of a clock shall be taken as positive, and angles measured clockwise shall be negative.*   With this understanding, we have such relations as

$$\angle ABC + \angle CBD = \angle ABD$$

regardless of the relative positions of the lines BA, BC, BD.

It is sometimes advisable to attach a sign to the distance from a point to a fixed line.   In this case the custom is to attach arbitrarily the positive sign to the distance to the line from every point on one side, the negative sign from the other.   It is customary to designate as positive the perpendiculars to the sides of a triangle from points within the triangle.

5. It is usual to regard an area as a signless or essentially positive quantity; but sometimes it is desirable to attach signs to areas.   When an area is determined as the product of two (directed) lengths, we may adopt as its sign the algebraic sign of this product.   Another method is to consider the two directions in which the boundary of an area may be described. If the circuit is traversed in a positive or counter-clockwise direction, the area shall be taken as positive; and if it is traversed in a clockwise direction, the area is negative. In our work, however, we shall hardly find these distinctions necessary.

6. Ratio of segments. In accordance with the foregoing, the ratio of two segments of a line will be positive or negative, according as the segments extend in the same or in opposite directions. We now consider two fixed points A, B, and any other point P on the line AB; we define the segments cut by P on AB as the signed distances $\overline{PA}$ and $\overline{PB}$; and the ratio in which P cuts AB, as $\overline{PA}/\overline{PB}$. We see that this ratio is independent, in magnitude and sign, of the choice of unit length and of positive direction on the line; and that it is negative for all positions of P between A and B, and positive when P is outside the segment AB. Let us now suppose a point P to trace the whole line through A and B, and consider the variation of the ratio

$$r = \frac{\overline{PA}}{\overline{PB}}$$

When P is far away in the direction BA, r is slightly less than one. As P approaches nearer to A, r passes through all values from one to zero; as P passes through A, and moves toward B, r becomes zero and then passes through all negative values, becoming -1 at the mid-point of AB. As P approaches B, r passes through numerically larger and larger negative values. After P has passed through B, the ratio is positive and very large; and finally, as P moves off in the direction of AB extended, the ratio decreases to the limiting value $+1$.

We see then that for every point of the line except B, a value of the ratio is determined; and conversely, that every value of r except $+1$ determines a point on the line. This may be formulated and proved algebraically:

7. Theorem. *If A and B are any two points and k any number different from $+1$, there exists one and only one point on the line for which the ratio $\overline{PA}/\overline{PB}$ equals k.*

Proof: let $$\overline{AB} = a, \quad \overline{PA} = x$$

then $$\overline{PB} = \overline{PA} + \overline{AB} = x + a$$

Then the desired relation is

$$\frac{\overline{PA}}{\overline{PB}} = \frac{x}{a+x} = k$$

Solving for x, we get

$$(1 - k) x = ak$$

which yields a unique value of x except, as aforesaid, when $k = 1$.

POINTS AT INFINITY

8. It frequently happens that a theorem deals with a point determined as the intersection of two lines of a figure. In the special case that the two lines in question are parallel, the theorem will have no meaning. In order to eliminate such exceptional cases, we adopt artificially a fiction that parallel lines have in common a point, which may be called a point at infinity.

Consider two lines, one of which is held fast, while the second is rotated about one of its points (other than the point of intersection). As the lines approach the position of parallelism, the point of intersection moves farther and farther off; when the lines become actually parallel, the point of intersection has disappeared. Thus arises the familiar phrase "parallel lines meet at infinity," which in the popular mind is invested with esoteric significance, as implying that there is an actual point, perhaps at a distance greater than can be conceived, where the parallels really meet. Of course, nothing could be farther from the fact; the proper statement is that the lines do not meet at all, and the phrase quoted is entirely meaningless.

For the very reason that the sentence has no intrinsic meaning, we are free to attach to it any meaning that we wish, and to use it with that interpretation. Such an extension of the meaning of a word or a phrase is a common custom in Mathematics,* and other instances will occur a little later. We have then the following definition.

9. Definition. The statement that two or more lines *intersect*, or *meet at a point*, or *are concurrent*, shall be interpreted as meaning either of two things: either there is a point through which all the lines pass, or else they are all parallel. Two or more parallel lines shall be said to have a *common point at infinity*, or to *intersect at infinity*.

A point at infinity, as such, we do not define; the statement that two or more lines have a common point at infinity shall merely be interpreted as another way of saying that the lines are parallel.

The most usual application of this definition is to the following situation. In a theorem a line is required to pass through the point of intersection of two given lines; if in a particular case these two lines become parallel, the other line shall be parallel to them. In other words, if a theorem asserts that three lines are concurrent, it shall be deemed to be true when the lines are parallel.

10. We may formulate and interpret the following supplementary definitions and propositions:

---

* "'That's a great deal to make one word mean,' Alice said in a thoughtful tone.
"'When I make a word do a lot of work like that,' said Humpty Dumpty, 'I always pay it extra.'"

a. All points at infinity lie on a straight line called *the line at infinity.*

b. If P is the point at infinity on line AB, then

$$\overline{PA}/\overline{PB} = 1 \qquad\qquad (cf.\ \mathbf{6},\ \mathbf{7})$$

(Let the reader state carefully the interpretation of each of these definitions, and establish the following propositions.)

c. On any line there is one and only one point at infinity.

d. Two lines in a plane determine a single point.

e. Two points in a plane, finite or infinite, determine a single line.

11. As a simple instance of the value of this point of view, let us recall a theorem from plane geometry.

Theorem. *The bisector of an exterior angle of a triangle divides the opposite side externally in the ratio of the adjacent sides.*

This theorem ceases to have any meaning when the adjacent sides are equal, unless we adopt some such special convention as that just established. With this convention it remains valid, for the bisector of the angle meets the opposite side at infinity, and therefore divides it in the ratio $+1$, which is precisely the ratio of the adjacent sides. Again, it can be proved that the three points where the bisectors of the exterior angles of a triangle meet the opposite sides are in general collinear. By virtue of our conventions, this statement is equally valid for an isosceles or even an equilateral triangle. The reader will study this situation carefully, and verify these conclusions.

12. We introduce at this time certain other generalizations, and extensions of definitions, which will be useful. In general, a circle can be drawn through three points, but an exception arises when the points are on a straight line.

In order to reduce the two cases to one, we extend the definition of the word *circle* so as to admit straight lines to the fellowship of circles. That is, by the word *circle* we may mean either a proper circle, with center and radius, according to the usual definition, or a straight line. In the latter case, the center is represented by the point at infinity in the direction perpendicular to the line; and the reciprocal of the radius is replaced by the value zero. It is permissible, and sometimes useful (as in **118**) to treat as a circle the combination of a straight line and the line at infinity.

Another special or limiting type of circle is the null-circle, whose radius is zero. We agree that any point may be regarded as a circle whose center is at the point, and whose radius is zero. The wording of a theorem, and the context, will usually make it clear whether we are restricted to proper circles, or whether either special type may be admitted to consideration.

NOTATION

13. In the study of the triangle, a standardized notation conduces to clearness and economy of statement. We shall deal always with a scalene triangle unless otherwise specified. (The modification of a theorem that is often necessary when the triangle becomes a right or an isosceles triangle will usually be obvious without specific formulation.)

Let A_1, A_2, A_3 designate the vertices of the triangle;

a_1, a_2, a_3, the lengths of the sides $\overline{A_2A_3}$, $\overline{A_3A_1}$, $\overline{A_1A_2}$;

α_1, α_2, α_3, the angles;

O the center and R the radius of the circumscribed circle;

O_1, O_2, O_3, the feet of the perpendiculars on the sides from O, the mid-points of the respective sides A_2A_3, A_3A_1, A_1A_2;

H_1, H_2, H_3, the feet of the altitudes, perpendicular to the sides from the opposite vertices A_1, A_2, A_3;

H the orthocenter, or point of intersection of the altitudes;

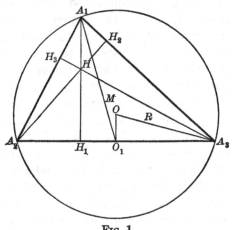

FIG. 1

h_1, h_2, h_3, the lengths of the altitudes;

m_1, m_2, m_3 the lengths of the medians $\overline{A_1O_1}$, $\overline{A_2O_2}$, $\overline{A_3O_3}$;

M the median point, or intersection of the medians;

s the half sum of the sides;

Δ the area;

X_1, X_2, X_3, when associated with a point X, usually the feet of the perpendiculars on the sides from X;

I the center and ρ the radius of the inscribed circle;

letters will be assigned to other points as they are introduced.

Exercise. *Prove that the perpendicular bisectors of the sides of a triangle meet at a point; also the angle bisectors; also the medians; also the altitudes.*

(Proofs of these theorems are to be found in any Plane Geometry; see also **219**.)

14. Drawing. It is highly desirable that the reader make complete and accurate drawings of all important theorems. Such drawings, especially in cases where the figure is at all complicated, are far more instructive than printed diagrams. The student should provide himself with ruler and compasses and with a pair of draughtsman's triangles; and should learn the use of these triangles for the construction of parallels and perpendiculars. The only constructions required in this book are those which can be effected with ruler and compasses, but much time is saved by the use of triangles or other special instruments for the frequently recurring constructions. (Where a mere approximation can be as accurate as the standard method of construction, the former should be used. For example, to draw a tangent to a circle from an outside point, or a common tangent to two circles, the best and most economical method is to lay the ruler carefully in the position of tangency, and at once draw the line. The point of tangency is found by laying a triangle on the ruler and locating the foot of the perpendicular to the tangent from the center of the circle.)

In the study of the triangle, frequent drawings will be necessary. It will be found rather difficult to draw offhand a triangle which is neither right-angled nor isosceles. The best procedure is to draw first the circumscribed circle, and to inscribe the triangle in this circle. We can control the shape of the triangle by means of the distances of the sides from the center of the circle.

Exercise. In a circle of about three inches radius, draw a triangle $A_1A_2A_3$, whose sides A_2A_3 and A_3A_1 are respectively about $\frac{3}{4}$ inch and $1\frac{1}{2}$ inch from the center. This will be a scalene triangle. Using triangles, draw the perpendiculars OO_1, OO_2, OO_3 to the sides from the center O of the circle. Draw also the altitudes A_1H_1, A_2H_2, A_3H_3, and determine their point of intersection H. Find also the point M of intersection of the medians A_1O_1, A_2O_2, A_3O_3.

Extend OO_1, OO_2, OO_3 to meet the arcs of the circle at

P_1, P_2, P_3; and draw the angle bisectors A_1P_1, A_2P_2, A_3P_3. Thus locate the incenter I, and draw the inscribed circle.

Find the center of the circle through H_1, H_2, H_3; and draw this circle.

15. The following theorems of the triangle may be established without difficulty. Formula a is deduced at once from the figure, and from it we derive b and d directly. Equation c is a combination of two or three standard theorems of elementary geometry, including the Pythagorean theorem, and also appears as the trigonometric law of cosines. Equations e and f are usually derived in geometry texts, while g is a simple combination of c and d.

a. $\quad h_1 = a_2 \sin \alpha_3 = a_3 \sin \alpha_2$

b. $\quad \dfrac{a_1}{\sin \alpha_1} = \dfrac{a_2}{\sin \alpha_2} = \dfrac{a_3}{\sin \alpha_3} = 2R \qquad$ *(law of sines)*

c. $\quad a_1{}^2 = a_2{}^2 + a_3{}^2 - 2\,a_2 a_3 \cos \alpha_1 \qquad$ *(law of cosines)*

d. $\quad \Delta = \tfrac{1}{2}\, a_1 h_1 = \tfrac{1}{2}\, a_2 a_3 \sin \alpha_1$

$$= 2\,R^2 \sin \alpha_1 \sin \alpha_2 \sin \alpha_3 = \frac{a_1 a_2 a_3}{4\,R} = \rho\, s$$

e. $\quad h_1 = \dfrac{2}{a_1} \sqrt{s\,(s - a_1)\,(s - a_2)\,(s - a_3)}$

f. $\quad \Delta = \sqrt{s\,(s - a_1)\,(s - a_2)\,(s - a_3)}$

g. $\quad \cot \alpha_1 = \dfrac{a_2{}^2 + a_3 - a_1{}^2}{4\,\Delta}$

It is to be borne in mind that any proposition in which the subscripts do not appear symmetrically may be repeated with the subscripts advanced in cyclic order, and thereby be made to apply to any of the sides and angles of the triangle.

Exercise. Give complete proofs of formulas a–g.

DIRECTED ANGLES

16. We now proceed to a specialization of the definition of angle in a form that is found to be highly useful throughout the work.* The symbol $\angle$ will be used, as hitherto, to denote the angle defined in the usual way, while the directed or oriented angle about to be defined will be designated by the symbol $\measuredangle$.

Definition. The *directed angle from a line l to a line l'*, denoted by $\measuredangle l, l'$, is that angle through which l must be rotated *in the positive direction* in order to become parallel to, or to coincide with l'. Similarly, $\measuredangle ABC$, *the directed angle from AB to BC*, is that angle through which the line AB, taken as a whole, must be rotated about B in the positive direction in order to coincide with BC.

From the definition it follows that directed angles are equivalent when they differ by 180° or multiples thereof. The magnitude of $\measuredangle ABC$ may be equal to that of $\angle ABC$, or of its supplement. If ABC is a triangle described in the positive or anticlockwise direction, it may be seen that the directed angles $\measuredangle ABC$, $\measuredangle BCA$, $\measuredangle CAB$ are equal to the respective exterior angles of the triangle, while the interior angles are $\measuredangle CBA$, $\measuredangle ACB$, $\measuredangle BAC$.

17. Addition of directed angles is defined as follows:

$$\measuredangle l_1, l_2 + \measuredangle l_2, l_3 = \measuredangle l_1, l_3$$
$$\measuredangle l_1, l_2 + \measuredangle l_3, l_4 = \measuredangle l_1, l_5$$

where l_5 is a line such that $\measuredangle l_2, l_5 = \measuredangle l_3, l_4$

18. As immediate consequences of these definitions we have the following rules of operations with directed angles:

Theorems.

a. $\measuredangle l_1, l_2 + \measuredangle l_2, l_1 = 0 \text{ or } 180°$

b. If l_1 is parallel to l_1' and l_2 to l_2', then

$$\measuredangle l_1, l_2 = \measuredangle l_1', l_2'$$

and similarly if the corresponding lines are perpendicular.

---

* Johnson: "Directed Angles in Elementary Geometry," *American Mathematical Monthly*, vol. XXIV, 1917, p. 101; "Directed Angles and Inversion, with a proof of Schoute's Theorem," *ibid.*, p. 313; "The Theory of Similar Figures," *ibid.*, vol. XXV, 1918, p. 108.

c. For any four lines we have the following identity:

$$\angle\, l_1,\, l_2 + \angle\, l_3,\, l_4 = \angle\, l_1,\, l_4 + \angle\, l_3,\, l_2$$

For
$$\angle\, l_1,\, l_2 = \angle\, l_1,\, l_4 + \angle\, l_4,\, l_2$$

and
$$\angle\, l_3,\, l_4 = \angle\, l_3,\, l_2 + \angle\, l_2,\, l_4$$

d. Three points A, B, C are collinear if and only if

$$\angle\, ABC = 0$$

or, an alternative and equivalent, if for any fourth point D,

$$\angle\, ACD = \angle\, BCD$$

e. If AB, AC are the equal sides of an isosceles triangle ABC, then

$$\angle\, ABC = \angle\, BCA$$

and conversely.

f. For any four points A, B, C, D,

$$\angle\, ABC + \angle\, CDA = \angle\, BAD + \angle\, DCB$$

19. We come now to the fundamental theorem of directed angles, which lends to the method its great usefulness.

Theorem. *Four points A, B, C, D, lie on a circle if and only if*

$$\angle\, ABC = \angle\, ADC$$

Let us recall that angles inscribed in the same arc are equal, and conversely. Also, angles inscribed in opposite arcs of a circle are supplementary; in other words, the opposite angles of a convex quadrilateral are supplementary if and only if the vertices are concyclic. These two well-known theorems may be amalgamated into one general theorem, namely: if four points A, B, C, D are on a circle, the angles $\angle ABC$ and $\angle ADC$ are equal or supplementary, accord-

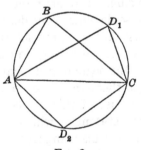

Fig. 2

ing as B and D are on the same side of BC, or on opposite sides; and conversely. Now let us note that if B and D are on the same side of AC, with $\angle ABC = \angle ADC$, then it follows that $\measuredangle ABC = \measuredangle ADC$; and conversely. But if B and D are on opposite sides of AC, and $\angle ABC$ and $\angle ADC$ are supplementary, still $\measuredangle ABC = \measuredangle ADC$; and conversely. This establishes the theorem for the cases when the points are not collinear; and finally, we have already seen that if and only if

$$\measuredangle ABC = \measuredangle ADC = 0$$

will the four points be on a line. Hence in all cases the equation

$$\measuredangle ABC = \measuredangle ADC$$

is a necessary and sufficient condition that the four points lie on a circle (in the extended sense of the word).

Corollary. *If A and B are fixed points, the locus of a point P for which $\measuredangle APB$ has a constant value is a circle through A and B.*

In conclusion, it may be remarked that the use of directed angles is by no means essential to the study of geometry. The method will be found to simplify and clarify to a remarkable degree many of our theorems and proofs, so that a single statement and proof will cover a situation that otherwise demands consideration of numerous cases, with less concise and exact formulations. The following chapters use the system freely; a good example of its advantages will be found at the beginning of Chapter VII, where an important and simple theorem is proved both by the usual methods and with the use of this device. However, we can always translate any statement which is in terms of directed angles back into the familiar language, merely by remembering that when two directed angles are asserted to be equal, the angles in the figure bounded by the same lines are actually equal or supplementary, according to the direction in which they are described. But it is this very uncertainty, and the consequence that in different figures the situation will be different, that impels us to introduce directed angles. Under this system it is a matter of indifference how the angles are actually related to each other; and the conclusions follow, independent of the accidental variations in the arrangement of the figure.

CHAPTER II

SIMILAR FIGURES

20. This chapter is devoted to a study of the relations of two similar figures in a plane. We recall that it is proved in elementary geometry that if all corresponding angles * of two figures are equal, all corresponding lines are proportional and the figures are similar. We shall discuss first two similar figures whose corresponding sides are parallel, and shall prove that the lines through all pairs of corresponding points are concurrent at a point called the homothetic center of similitude. In the more general case, when two similar figures lie in a plane but corresponding sides are not parallel, there exists a center of similitude, or self-homologous point, which occupies the same homologous position with reference to the two figures. The properties of this point are developed below in sufficient detail to be useful later, and the special case of two circles is given the attention that it merits.

21. We consider first the problem of similar figures in what is called homothetic position, namely with homologous lines parallel and with the connectors of homologous points concurrent (Fig. 3).

Theorem. *Let a point O and a figure $ABC \ldots$ be given, and let each of the lines OA, OB, $OC \ldots$ be divided in a constant ratio k at A', B', C'. . . . Then the figure $A'B'C' \ldots$ is directly similar to $ABC \ldots$ with corresponding sides parallel.*

For we see at once that any two corresponding triangles, such as OAB and $O'A'B'$, are similar; therefore corresponding lines are parallel, and are proportional in the ratio k.

---

* That is, the angles of all triangles formed by drawing all possible corresponding lines.

It may be remarked that the given figure need not be rectilinear. For example:

Theorem. *The locus of the mid-points of the lines connecting a fixed point with the points of a circle is a second circle of half the radius of the first. Corresponding radii are parallel, and if tangents are drawn at corresponding points they are parallel.*

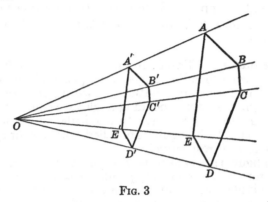

Fig. 3

22. As another extension, we may note that the point O of the first theorem need not lie in the plane of the figure ABC.... We may formulate the following theorems.

Theorem. *If O is a point outside the plane of the figure ABC ... and if each of the lines OA, OB, OC be cut in a given ratio k at A', B', C', ... then A', B', C' ... lie in a plane parallel to ABC ... and the two figures are similar. Conversely, any plane parallel to the given plane cuts these rays in a figure similar to the given figure; and again, if similar figures lie in parallel planes with sides parallel, the connectors of corresponding points are concurrent.*

23. Continuing the discussion of the figures in a single plane, we see that if any two homologous points are connected, it is easily proved by similar triangles that the connector passes through O. This point is called the *homothetic center*, or *center of similitude*, and the ratio k is called the *ratio of similitude* of the two figures. This ratio may have any value, positive or negative; if it is positive, homologous points are on the same side of O and homologous lines are drawn in the same direction, while if it is negative, O separates each pair of homologous points and corresponding lines extend in opposite directions. In the former case, the center of similitude is said to be *external*, and in the latter case *internal*.

24. Definitions. Two figures are said to be *directly similar*, when all corresponding angles are equal and are described in the same sense of rotation. Two figures are *inversely similar* when all corresponding angles are equal but are described in opposite senses. In particular, if corresponding lines are equal, the figures are called *congruent* or *symmetrically congruent* respectively.

Two figures are said to be *homothetic* to each other, if the connectors of corresponding points are concurrent at a point which divides each connector in the same ratio. The relation between homothetic figures is called *expansion*.

Theorem. *Conversely, if two figures are similar, with corresponding sides parallel, they are homothetic; that is, there is a center of similitude O, through which pass the connectors of all pairs of corresponding points.*

The proof by means of similar triangles is immediate. As before, there are two cases, according as homologous parallel lines extend in the same or in opposite directions. An exceptional case arises when the figures are congruent with homologous lines extending in the same direction, for then all connectors of corresponding points are parallel, and the center of similitude is at infinity.

25. *Two circles* may be regarded as homothetic to each other in two ways.

Theorem. *If in two non-concentric circles we draw radii parallel and in the same direction, the line connecting their extremities passes through a fixed point on the line of centers, which divides that line externally in the ratio of the radii. If parallel radii are drawn in opposite directions, the line joining their extremities passes through the point which divides the line of centers internally in the ratio of the radii.*

Definition. These two points, which divide the line of centers of two circles externally and internally in the ratio of the radii, are called respectively *the external and internal center of similitude* of the circles, or *the external and the internal homothetic center.*

Corollary. *If the two circles have direct common tangents, these pass through the external center of similitude; and if the transverse common tangents exist, they pass through the internal center of similitude.*

Corollary. *From a point of intersection of two circles, the lines to the centers of similitude bisect the angles between the radii of the circles.*

26. Homologous and antihomologous points. The extremities of parallel radii of two circles are called *homologous* with regard to the center of similitude collinear with them. Two points that are collinear with a center of similitude, but are not homologous, are said to be *antihomologous* with regard to that center of similitude.

That is, a line through a center of similitude cuts the circles in four points; to either point on one circle one of the points on the other circle is homologous, the other is antihomologous.

The concept of homologous and antihomologous points on two circles is a decidedly useful one, and we pause to develop it in somewhat fuller detail.

27. Theorem. *Two pairs of antihomologous points form inversely similar triangles with the homothetic center. That* *is, if with regard to a center of similitude C, P and Q are antihomologous respectively to P' and Q', then triangles CPQ and CQ'P' are inversely similar.*

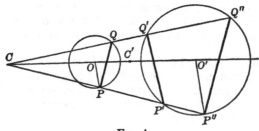

Fig. 4

Let O, O' be the centers of the circles; let two lines through the center of similitude C be $CPP'P''$ and $CQQ'Q''$, so that $O'P''$ and $O'Q''$ are respectively parallel to OP and OQ; then P' and Q' are antihomologous to P and Q respectively. Evidently PQ and $P''Q''$ are parallel, arcs PQ and $P''Q''$ are similar and measure equal angles. Hence

$$\measuredangle\, CPQ = \measuredangle\, CP''Q'' = \measuredangle\, P'Q'Q'' = \measuredangle\, P'Q'C,$$
$$\measuredangle\, PQC = \measuredangle\, P''Q''C = \measuredangle\, P''P'Q' = \measuredangle\, CP'Q',$$

showing that the triangles are inversely similar, since corresponding elements are in reverse order.

28. Theorem. *The product of the distances from a homothetic center to two antihomologous points is constant.*

For in the above figure, we have in the similar triangles

$$\overline{CP} \cdot \overline{CP'} = \overline{CQ} \cdot \overline{CQ'}$$

and if P moves, while Q is held fast, the product $\overline{CP} \cdot \overline{CP'}$ remains constant.

29. Theorem. *Any two pairs of points, antihomologous with regard to a center of similitude, are on a circle.*

For the equations in the proof of **27** are equivalent to

$$\angle P'PQ = \angle P'Q'Q$$

which shows (**19**) that P, P', Q, Q' are on a circle. This can be simply established later in another way, by virtue of **42**.

30. Theorem. *The tangents to two circles at antihomologous points make equal angles with the line through the points.*

For the tangents at P and at P'' are parallel, while those at P'' and P' make equal angles with PC.

Corollary. *The line joining two antihomologous points, and the tangents at those points, form an isosceles triangle. Conversely, if two equal tangents are drawn to two circles from an outside point (**45**) the points of contact are antihomologous.*

SIMILAR FIGURES IN GENERAL

31. We now proceed to consider the more general question of the relations to each other of two directly similar figures in a plane. We recognize four fundamental operations on a figure, associated with the concept of similarity, namely:

a. translation, or a sliding motion whereby every point of the figure moves the same distance in the same direction;
b. rotation of the figure about a fixed point;
c. expansion with regard to a fixed center of similitude (**24**);
d. reflection with regard to a line, which is the same as turning the figure over on this line as an axis.

It is obvious that if these operations are performed any number of times on a figure, the resulting figure will be similar to the original, directly or inversely according as the number of reflections is even or odd; and the ratio of similitude of the last to the first figure is precisely the product of the ratios of the expansions that have taken place. By rotations and translations, a figure is carried into a congruent figure; and by a reflection, into a symmetrically congruent figure. Conversely, if two figures are similar, they can be made to coincide by a series of these operations; for example, by a translation that causes a point A to fall on the corresponding point A', followed by a rotation and an expansion. Our present purpose is to reduce this last statement to its lowest terms; and our result will be that in general two directly similar figures can be made to coincide by a combination of an expansion with respect to a certain point, and a rotation about the same point.

32. Definition.   For the time being, we will use the word *homology* to designate a rotation about a point, accompanied by an expansion with regard to that point. As special cases, mere rotation and mere expansion are included in the definition.

Theorem.   *Any segment AB can be carried into any segment $A'B'$ in one and only one way by an operation which is either a translation or a homology.*

Special case 1: If $AA'B'B$ is a parallelogram, a translation along AA' and BB' achieves the result.

Special case 2: If AB and $A'B'$ are parallel, but AA' and BB' intersect at C, a suitable expansion with regard to C as homothetic center will bring AB to fall on $A'B'$.

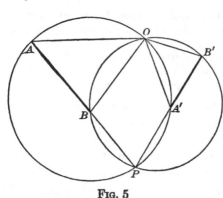

Fɪɢ. 5

General case: Let AB intersect $A'B'$ at P, and let none of the four points coincide with P. Let the circles through A, A', P and B, B', P intersect again at a point O. We see at once that triangles OAB and $OA'B'$ are directly similar; for $\angle OAB = \angle OA'P = \angle OA'B'$, etc.

Hence if we rotate OAB about O until A falls on OA', and then expand with regard to O as center until A falls on A', the line AB will fall on the line $A'B'$.

If one of the given points, say B, coincides with P, we use instead of the circle BPB' the circle through B' which is tangent to AB at B. The proof then holds without modification. Again, it may happen that the two circles are tangent at P, so that O and P coincide; which means that AA' and BB' are parallel.

Theorem. *If two figures are known to be directly similar, and two points of one coincide with the homologous points of the other, the two figures coincide throughout.*

In itself this is trivial, but in combination with the preceding theorem, it leads to our principal result, namely:

33. Theorem. *If two figures are congruent, there exists either a single rotation or a single translation that brings the one to coincide with the other. If two figures are directly similar but not congruent, there exists a unique homology that transforms the one into the other.*

Definition. The center of the homology or the rotation is called the *center of similitude* of the figures; the angle of rotation and the ratio of expansion are called the *angle of similitude* and the *ratio of similitude* respectively.

Construction. To locate the center of similitude, we draw circles, each passing through a pair of homologous points and the intersection of any homologous lines through them. Every such circle passes through the center of similitude.

34. If any point has the properties of a center of similitude, it is the point O thus determined; therefore the center of similitude for transforming the first figure to the second is also the center of similitude for the reverse operation.

Theorem. *In two directly similar figures, the center of similitude is homologous to itself; and conversely a point which is self-homologous is the center of similitude.*

This theorem often makes it easy to determine the center of similitude of two figures.

35. As an interesting interpretation of the foregoing theory, we may consider the properties of maps. If two maps of the same plane region are superposed, the same side up, on a plane, then by our theorems there will be one and only one point whose representations in the maps will coincide. If the maps are on the same scale, either may be rotated about this point so that they may be made to coincide; except in the one case that the self-corresponding point is at infinity, when one map can be made to coincide with the other by moving it parallel to itself. If they are on different scales, the self-corresponding point is necessarily a finite point (though of course it may fall far outside the boundaries of the actual maps); and the triangle formed by this *self-corresponding* or double point with any pair of corresponding points is of constant form.

36. A few illustrations and applications follow. In later chapters we shall have frequent occasion to make use of the center of similitude or self-homologous point of similar figures.

a. **Theorem.** *If a triangle is of fixed form, and one vertex remains fixed, while a second traces any figure, the third traces a similar figure, with the center of similitude at the fixed point.*

b. **Theorem.** *If two directly similar figures are inscribed in the same circle, they are congruent, and their center of similitude is the center of the circle.*

c. **Theorem.** *The triangle whose vertices are at the midpoints of the sides of a given triangle is homothetic to the latter, with ratio of similitude − ½ and with the center of similitude at the median point M.*

d. **Corollary.** What conclusion is drawn from the fact that the altitudes of the smaller triangle are concurrent at the center of the circumscribed circle of the original triangle?

37. From the point of view of **31**, two circles may be regarded as similar to each other in an infinite number of ways, with any points on the respective circles chosen as homologous. More definitely, we may think of similar polygons inscribed in the circles; and as one of the polygons rotates about the center of its circle, the center of similitude will occupy various positions, whose locus we now seek. Evidently in any position of the two figures, the distances from the center of similitude to the centers of the circles are in a constant ratio, that of the radii of the circles. Hence if the circles are equal, the locus in question is the perpendicular bisector of the line of centers.

Theorem. *The locus of the centers of similitude of two non-concentric circles is a circle having as diameter the line joining the two homothetic centers* (**25**).

Obviously the locus passes through the homothetic centers. Let the circles be $O(r)$ * and $O'(r')$, the homothetic centers E and I; and let P be any position of the center of similitude. Now we know that

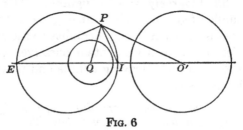

$$\frac{\overline{OP}}{\overline{O'P}} = \frac{r}{r'}$$

Fig. 6

But E and I divide the segment OO' externally and internally in the ratio r/r'; hence in triangle POO', PE and PI divide OO' into segments proportional to PO and PO', and are therefore the bisectors of the angles at P. But the bisectors of supplementary adjacent angles are necessarily perpendicular to each other. Hence EPI is a right angle, and P lies on a circle with EI as diameter, as was to be proved.

* That is, the circle whose center is O and radius r.

In the special case of equal circles, E is at infinity and the circle becomes a straight line, as above indicated.

Definition. The circle whose center lies on the line of centers of two given circles, and whose circumference divides that line internally and externally in the ratio of the radii, is called the *circle of similitude* of the circles.

Each point of this circle is the center of a homology with regard to which the given circles correspond. Therefore the circle is the locus of a point whose distances from the centers are proportional to the radii, and also the locus of a point whence the circles subtend equal angles. If the given circles are equal, the circle of similitude is a straight line; if either circle is a line, the circle of similitude does not exist; if either is a null-circle, but not both, the circle of similitude coincides with the null-circle. Two concentric circles have a single center of similitude, their common center.

38. Inversely similar figures. The theory of inversely similar figures which is analogous to the foregoing is by no means so general in applicability to the problems of geometry. For that reason we content ourselves with a mere outline of the main facts, leaving details and proofs to the reader who is interested.*

Given two symmetrically congruent figures in a plane, there exists a definite axis, such that either figure can be made to coincide with the other by a reflection with regard to this axis, together with a translation along the same axis.

In terms of maps, this is to say that if two copies of the same map are superposed, facing each other, there is always a line having the property that one map can be folded over on this line as axis and then slid along the same line, so that it will come to coincide with the other.

* A fairly satisfactory discussion is given by Lachlan, *Modern Pure Geometry*, p. 134.

Given two inversely similar figures, not on the same scale, there exist two mutually perpendicular axes of similitude, whose point of intersection is called the center of similitude. If one figure is reflected with regard to either axis, and then subjected to a homology with regard to this center, it comes to coincidence with the other.

Exercise. *Prove that if two figures are inversely similar, the bisectors of the angles between corresponding lines are parallel to two fixed directions, the directions of the axes of similitude.*

Exercise. *Prove and amplify the statement that the net effect of two successive reflections is the same as that of a rotation about the point of intersection of the axes. What is the angle of the rotation? Is it possible to choose two axes with regard to which the combined reflections will be equivalent to a given rotation?*

Exercise. Give complete proofs of the following propositions of this chapter, which are left unproved in whole or in part: **21, 22, 24, 25** (theorem and two corollaries), **30, 36** (four parts), **38.**

CHAPTER III

COAXAL CIRCLES AND INVERSION

39. This chapter is devoted to the study of systems of circles. We introduce first, by means of some familiar theorems of elementary geometry, the so-called power relation. The extended study of this relation leads us to the radical axis of two circles, and then to the properties of coaxal systems. We turn then to the very important theory of circular inversion and establish its essential principles. The present chapter contains so much of the theory as is needed in the study of the triangle, while in Chapter V some of the further developments of the same topics are carried out in more detail.

40. First of all, we reëstablish two standard theorems of elementary geometry, which we recast and combine into a single theorem in a form suited to our purposes.

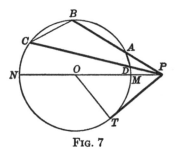

FIG. 7

Theorem. *If lines are drawn from a fixed point to intersect a fixed circle, the product of the distances from the fixed point to the points of intersection is constant.**

* In the geometry texts, this theorem is usually separated into cases, according as the fixed point is inside or outside the circle, and stated in such forms as the following: (a) if two chords of a circle intersect, the segments

For let P be the given point, either inside or outside the circle. Let PAB and PCD be two lines meeting the circle at A, B and C, D respectively. Then in every case, homologous angles of triangles PAD and PBC are measured by the same arcs; the triangles are similar, and

$$\overline{PA} \cdot \overline{PB} = \overline{PC} \cdot \overline{PD}$$

as was to be shown.

Corollary. *If the radius of the circle is r, its center at O, the constant product $\overline{PA} \cdot \overline{PB}$ is equal to*

$$p = \overline{OP}^2 - r^2$$

Definition. The *power* of a point P with regard to a circle with center O, radius r, is the quantity $\overline{OP}^2 - r^2$.

41. The reader will establish the following properties of the power, which depend directly on the foregoing theorem and definition.

Theorem. *If the point P is outside the circle, its power with regard to the circle is positive, and equal to the square of the tangent to the circle from P. If P is on the circle, the power is zero. If P is within the circle, the power is negative; it may be interpreted as the product of the segments of the diameter through P: $(\overline{OP} + r)(\overline{OP} - r)$, or as minus the square of the half-chord through P perpendicular to OP.*

The chord of a circle through an interior point P, perpendicular to OP and therefore bisected at P, is called the *minimum chord* for P, being the shortest chord that can be drawn through P. Certain power relations are neatly expressed in terms of the tangent to the circle when the point lies outside, and in terms of the minimum chord when it is inside.

are inversely proportional; (b) if a tangent and a secant are drawn to a circle, the tangent is a mean proportional between the whole secant and the external segment.

42. Theorem. *Conversely, if two lines AB and CD meet at P, and if*

$$\overline{PA} \cdot \overline{PB} = \overline{PC} \cdot \overline{PD}$$

not merely numerically but algebraically, then A, B, C, D are concyclic.

For by **40** a circle through three of the points must pass through the fourth; if the circle through A, B, C cuts PC at D', then $\overline{PA} \cdot \overline{PB} = \overline{PC} \cdot \overline{PD'}$; hence D and D' coincide.

43. For the sake of completeness, we define the power of a point with regard to a *null-circle* as the square of the distance from the point to the null-circle. When a circle degenerates into a *straight line*, the power of a point cannot be satisfactorily defined. (It is easily established, however, that in this case the *ratio of the power to the diameter* of the circle approaches, and in the limit is represented by the perpendicular from the point to the line.

For let O be the center of a circle, as before; r its radius, and let a line OP meet the circle at M and N. Then

$$p = \overline{OP}^2 - r^2 = \overline{PM} \cdot \overline{PN}$$

$$\frac{p}{2\,r} = \frac{\overline{PN}}{\overline{MN}}\,\overline{PM}$$

If now M remains fixed, while O and therefore N recede indefinitely on the line $PNOM$, the limiting form of the circle is that of the straight line through M perpendicular to PM. But the limit of the ratio $p/2\,r$ is obviously PM. Hence, in dealing with power theorems, if the limiting case where a circle reduces to a straight line is to come into consideration, we must state our theorems in terms of the ratio of power to diameter.)

44. Theorem. *The locus of a point having power k with regard to a fixed circle of a radius r is a concentric circle of radius $\sqrt{r^2 + k}$, provided that $r^2 + k > 0$.*

45. The fundamental theorem concerning the power relation is the following:

Theorem. *The locus of a point having equal power with regard to two given non-concentric circles is a certain line perpendicular to their line of centers. If the circles intersect, this is the line through their points of intersections.*

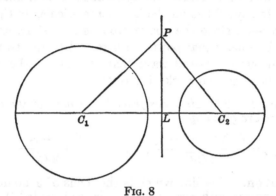

Let the circles be C_1 (r_1), C_2 (r_2). Let P be any point whose powers with regard to the circles are equal; and let PL be the perpendicular from P to the line of centers C_1C_2. Then, denoting $\overline{C_1L}$ by d_1, $\overline{C_2L}$ by d_2, $\overline{C_1C_2}$ by d, we have by hypothesis

$$\overline{PC_1}^2 - r_1^2 = \overline{PC_2}^2 - r_2^2$$
$$\overline{PL}^2 + d_1^2 - r_1^2 = \overline{PL}^2 + d_2^2 - r_2^2$$

whence $\qquad d_1^2 - d_2^2 = r_1^2 - r_2^2$

But $\qquad d_1 - d_2 = d$

whence by division

$$d_1 + d_2 = \frac{r_1^2 - r_2^2}{d}$$

Solving these equations simultaneously, we find

$$d_1 = \frac{d^2 + r_1^2 - r_2^2}{2d} \qquad -d_2 = \frac{d^2 + r_2^2 - r_1^2}{2d}$$

By inspection, these results are independent of P; hence for all positions of P under the hypothesis, L is a fixed point. That is to say, P lies on the line perpendicular to C_1C_2 at L.

Conversely, since the above proof is reversible, any point on this line has equal power with regard to the two circles. It is obvious in several ways that this line is the common chord when the two circles intersect.

Corollary. *The distances from the centers of the circles to the line are given by*

$$\overline{C_1L} = \frac{\overline{C_1C_2}^2 + r_1^2 - r_2^2}{2\,\overline{C_1C_2}}, \quad \overline{C_2L} = \frac{\overline{C_1C_2}^2 + r_2^2 - r_1^2}{2\,\overline{C_2C_1}}$$

Definition. The line which is the locus of a point having equal power with regard to two circles is called the *radical axis* of the circles. The radical axis of two concentric circles is defined as the line at infinity.

Any two circles have a radical axis. That portion of the radical axis which is exterior to the circles is the locus of a point from which the tangents to the circles are equal; and the portion within the circles, if any, is the locus of a point whose minimum chords (**41**) in the two circles are equal. Such a point is the center of a circle whose common chord with either of the given circles is a diameter of itself.

46. Theorem. *The radical axes of three circles, taken in pairs, are concurrent.*

For, the point in which any two radical axes intersect has equal power with regard to all three circles, and therefore lies on the third radical axis. The theorem is evidently still valid in the various special cases, namely if one or more of the circles be null, or if two of them be concentric, or if their centers be collinear.

Definition. The point of concurrence of the radical axes of three circles is called the *radical center*. If it lies outside the circles, it is the single point from which tangents to them are equal.

47. Problem. *To construct the radical axis of two circles.*

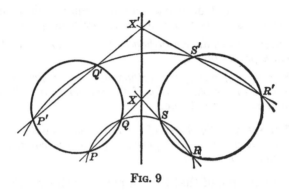

FIG. 9

If the given circles intersect at accessible points, the radical axis is the line through these points. If this is not the case, we draw any convenient circle, cutting one circle at P, Q and the other at R, S. Then the point of intersection of PQ and RS is the radical center of the three circles, and is a point on the desired radical axis. A second point may be found in like manner.

48. Definition. The angle between two intersecting circles is defined as the angle formed by the tangents at either point of intersection, or, equally, as the angle between the radii at either point.

A case of particular interest arises when the circles intersect at right angles.

Definition. Two circles are said to intersect *orthogonally*, or to be *mutually orthogonal*, when their angle of intersection is a right angle. In this case the tangent to either at a point of intersection passes through the center of the other.

A number of corollaries follow at once from the definition:

a. *If r_1 and r_2 are the radii of two circles, d the distance between their centers, the condition for orthogonality is*

$$r_1^2 + r_2^2 = d^2$$

b. *One circle and only one can be drawn orthogonal to a given circle, with its center at a given external point.*

c. *A circle can be drawn orthogonal to a given circle, through two given points thereof.*

d. *The center of one of two orthogonal circles lies on the circumference of the other, only if the former is a null-circle or the latter is a straight line.*

49. Theorem.   *The locus of the center of a circle orthogonal to two given circles is that portion of the radical axis which is exterior to the circles.*

For we have noted that from a point on this portion of the radical axis, the tangents to the circles are equal.

Corollary.   *Three circles have one and only one common orthogonal circle, whose center is at their radical center, provided this point lies outside the circles.*

Corollary.   *Any point of the common chord of two intersecting circles is the center of a circle, which has as diameters its common chords with each of the given circles; these are also the minimum chords of the latter (**41**). Similarly, if the radical center of three circles lies within the circles, their minimum chords through this point are diameters of a circle whose center is at the radical center.*

COAXAL CIRCLES

50. Definition.   A system of circles, each two of which have the same line as radical axis, is called a *coaxal system.*

We will consider the properties of such a system.

First, it is evident that the centers of the coaxal circles are on a line perpendicular to the common radical axis, say at a point L. It is then necessary and sufficient that L have the same power with regard to all the coaxal circles; if C (r) is one of these circles, $\overline{LC}^2 - r^2$ must be a constant for all the circles of any coaxal system.

51. *Case I*. Let this power be a positive quantity c^2. Then since $\overline{LC}$ is necessarily greater than r, no circle of the system meets the radical axis. We may assign to r any value and determine the center C; or we may assign to $\overline{LC}$ any value not less than c, and determine the value of r. Thus there is a circle of the system with any assigned radius, and there is one with its center at any point of the line of centers except the points within the segment KK', where $\overline{LK} = \overline{K'L} = c$. Let us consider the circle L (c) on KK' as diameter. It appears that every circle of the coaxal system is orthogonal to this circle, since $\overline{LC}^2 = c^2 + r^2$. Hence:

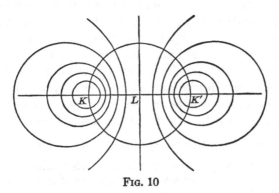

FIG. 10

Theorem. *A coaxal system of the first type consists of all circles whose centers are on a given line and which are orthogonal to a given circle whose center is on this line. The circles of the system can be drawn by drawing tangents to the given circle from points on the line of centers, and using these tangents as radii of the required circles. In a coaxal system of type I, there are two null-circles, the extremities K, K' of* *the diameter of the fixed circle. These points are called the limiting points of the coaxal system.*

52. *Case II.* If the constant power is a negative number $-d^2$, then each of the circles must cut the radical axis at a distance d on either side of L; for the equation

$$\overline{LC}^2 + d^2 = r^2$$

shows that the radius is the hypotenuse of a right triangle whose sides are LC and d.

Theorem. *A coaxal system of the second type consists of all circles through two fixed points.*

Every point on the line of centers is the center of one circle of the system, and with any assigned radius r there are two circles of the system, provided r is greater than d.

53. *Case III.* If the constant power is zero, we have evidently the system of circles tangent to the radical axis at L.

For the sake of completeness, we define a fourth type of coaxal system, namely that of all circles with a common center; the radical axis being the line at infinity. As a fifth type, the reason for whose inclusion will appear as we progress, we mention the set of straight lines through a point, including as a further specialization a set of parallel lines.

54. Summary. *Coaxal systems are of five types:*

  I. A set of non-intersecting circles whose centers are collinear and which are orthogonal to a fixed circle.

  II. A set of circles through two fixed points.

  III. A set of circles having a common tangent at a fixed point.

  IV. A set of circles with a common center.

  V. A set of concurrent lines.

Theorem. *Any two given circles are members of one and only one coaxal system.*

55. Theorem.   *A circle orthogonal to two fixed circles is orthogonal to each circle coaxal with them.*

For the center of such a circle is on the radical axis of the fixed circles; and its radius is equal to the tangent from the center to each circle.   But this tangent is the same for all circles coaxal with the fixed circles, and hence the same circle is orthogonal to them all.

Theorem.   *The circles orthogonal to two circles constitute a coaxal system.*

Corollary.   *Two given circles determine two coaxal systems, one consisting of all the circles coaxal with them, the second of all circles orthogonal to them.   Each circle of either system is orthogonal to every circle of the other; the radical axis of either system is the line of centers of the other.   If one system is of type I, the other is of type II, and the limiting points of the one are common to the circles of the second.   If either is of type III, the other is of the same type; if the one consists of concentric circles, the other consists of radiating lines; and finally, there may be two mutually perpendicular sets of parallel lines.*

Definition.   Two coaxal systems whose members are mutually orthogonal are called *conjugate.*

56. Problem.   *To construct the conjugate coaxal systems determined by two given circles.*

In the general case, it is sufficient to locate the points K, K' which are common to the circles of one system.   If the given circles do not intersect, any circle orthogonal to them intersects their line of centers at K and K'.   Then the one system consists of circles through these points; and tangents to any circle of this system, extended to meet the line KK', are radii of circles of the other system.

Problem.   *To construct the circle of a coaxal system which passes through a given point.*

Theorem.   *If a circle belongs to neither of two conjugate coaxal systems, there is not more than one circle of either system orthogonal to it.*

The proof, and the construction of the circle, depend on the determination of the radical center of the circle with either of the coaxal systems.

Theorem.   *There are in general two members of a coaxal system tangent to a given line or to a given circle.*

The construction involves the circle of the conjugate system which is orthogonal to the given line or circle.

57. Theorem.   *In a coaxal system of the second type, each circle is the locus of a point from which the common chord of the circles subtends a constant angle.*

That is, if we have the system of circles through two points K, K', then as a point P moves on any chosen circle of the system, the angle KPK' is obviously constant.

There is a somewhat analogous theorem for the coaxal system of the first type, for which we now establish some preliminary theorems.

58. The following important locus theorem is based on standard theorems of elementary geometry, and sometimes appears as an exercise in the school texts.

Theorem.   *If a point moves so that its distances from two fixed points are in a constant ratio, the locus is a circle whose center is collinear with the given points.*

This theorem has already been proved for a special case

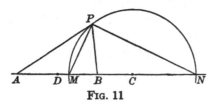

FIG. 11

(**37**).   If the constant ratio is 1, the locus is obviously the perpendicular bisector. Otherwise, let A, B be the given points, and let P be such a point that $\overline{PA}/\overline{PB}$ has a given value k different from 1. Let PM and PN bisect the angles formed by AP and BP; then M and N divide AB in the ratios $-k$ and k respectively. Hence as P changes its position, M and N are fixed points. But since MPN is a right angle, it follows that P describes a circle on the segment AB as diameter.

The following corollaries are established without difficulty:

a. $$\overline{MA} = -\frac{k}{k+1}\,\overline{AB} \qquad\qquad \overline{MB} = \frac{1}{k+1}\,\overline{AB}$$

$$\overline{NA} = \frac{k}{1-k}\,\overline{AB} \qquad\qquad \overline{NB} = \frac{1}{1-k}\,\overline{AB}$$

b. *The radius of the circle is*

$$r = \frac{k}{k^2-1}\,\overline{AB}$$

c. *The distance from A to its center C is*

$$\overline{AC} = \frac{k^2}{k^2-1}\,\overline{AB}$$

d. *The distance from the mid-point D of AB to the center of the circle is*

$$\overline{DC} = \frac{k^2+1}{2\,(k^2-1)}\,\overline{AB}$$

e. *The tangent to the circle from D is precisely $\tfrac{1}{2}\overline{AB}$, whatever the value of k.*

For $$t^2 = \overline{DC}^2 - r^2 = \frac{k^4 - 2\,k^2 + 1}{4\,(k^2-1)^2}\,\overline{AB}$$

f. *Therefore whatever the value of k, the circle is orthogonal to the fixed circle on AB as diameter; in other words, the circles derived by assigning different values to k are members of a coaxal system having A and B as limiting points.*

59. Theorem. *Conversely, if A and B are two points which divide a diameter MN of a circle externally and in-*

ternally in the ratios $\pm k$, then the distances from a moving point P of the circle to A and B are in the constant ratio k.

The result can be stated also in the following form:

Theorem. *If one side of a triangle and the ratio of the other two sides are given, the locus of the third vertex is a certain circle whose center is on the extension of the given side.*

For a given triangle this theorem defines three circles, known as the *circles of Apollonius*. These will be studied in Chapter XVII.

The circle of similitude of two circles is a special case of the foregoing; and it appears that the circle of similitude is a member of the coaxal system whose limiting points are the centers of the given circles. That the circle of similitude is also coaxal with these circles can be most easily proved later (**115**).

60. The foregoing results lead us to a general theorem concerning coaxal circles, namely:

Theorem. *In a coaxal system of the first type, each circle is the locus of a point whose distances from the limiting points are in a constant ratio k.*

This was established in **58** f. It may be noted that in conjugate coaxal systems with basal points K, K', each circle of one system is the locus of points for which $\measuredangle KPK'$ is constant, while each circle of the other is the locus of points for which the ratio $\overline{PK}/\overline{PK'}$ is constant.

Corollary. *If a line segment is divided externally and internally in ratios numerically equal, the circle having the points of division as extremities of a diameter is a member of the coaxal system whose limiting points are the extremities of the given segment.*

Corollary. *The locus of a point from which two collinear segments AB, BC subtend equal angles is a circle passing through B.*

61. The following theorems establish some interesting associations between the homothetic centers and antihomologous points of two circles on the one hand, and the radical axis on the other.

a. **Theorem.** *If P and Q, and R and S, lying respectively on two circles, are antihomologous with regard to the same homothetic center, the secants PR and QS intersect on the radical axis.*

For we proved (**29**) that the four points lie on a circle; whence as in **47**.

b. **Theorem.** *If either homothetic center of two circles is known, the radical axis can be constructed with ruler only.*

c. **Theorem.** *Tangents to two circles at antihomologous points intersect on the radical axis.*

For by **30** two such tangents are equal.

d. **Theorem.** *Conversely, from a point on the exterior portion of the radical axis, four equal tangents to two circles can be drawn. The points of contact are antihomologous, two pairs with regard to either center of similitude of the circles. In other words, if a circle is orthogonal to two given circles, the points of intersection are antihomologous in this fashion.*

e. **Theorem.** *If radii are drawn to two circles at antihomologous points, and extended to intersect, their point of intersection is the center of a circle tangent to the given circles at the given points. Conversely, if a circle is tangent to two others, the points of contact are antihomologous, and the tangents at these points intersect on the radical axis.*

62. Exercises. Besides the following corollaries and exercises, the reader is to complete the proofs which have been omitted or merely sketched in the text, viz: **41, 44, 47, 48, 49, 51, 55, 56, 58, 59, 60, 61.**

a. **Theorem.** *On each side of a triangle, or its extension, a pair of points are marked, in such a way that each two pairs lie on a circle. Then all six points lie on one and the same circle.*

For if there were three distinct circles, their three radical axes, which are the sides of the triangle, would be concurrent. An alternative statement is:

b. **Theorem.**   *Given three lines, and on each a pair of points such that a circle passes through each two pairs; then either the three lines are concurrent or the six points lie on one circle.*

As a special case, if the points of any pair are coincident, the circle is tangent to the corresponding line.

c. **Theorem.**   *Two intersecting circles ABX and ABY are orthogonal if and only if*

$$\angle AXB + \angle AYB = 90°.$$

d. **Theorem.**   *If d is the distance between the centers of two circles, l their common chord, r and r' their radii, they are orthogonal if and only if*

$$ld = 2\,rr'.$$

e. **Theorem.**   *If a line is drawn through an intersection of two circles, meeting the circles again at P and Q respectively, the circles with centers at P and Q, each orthogonal to the other circle, are orthogonal to each other.*

f. **Theorem.**   *If AB is a diameter of a circle, and if any two lines AC and BC meet the circle again at P and Q, respectively, then the circle CPQ is orthogonal to the given circle.*

g. **Theorem.**   *The radical axes of a fixed circle with the members of a coaxal system are concurrent.*

h. **Theorem.**   *If two coaxal systems have a circle in common, they are orthogonal to a circle, or else cut a certain circle in diametrically opposite points* (cf. **49,** corollaries).

i. **Problem.**   *Investigate the properties of the system of circles which are orthogonal to one fixed circle.*

Henri Poincaré, in *Science and Hypothesis*, conceives a universe contained within a great sphere, in which the laws of temperature expansion and of refraction are such that the dimensions of objects vary as they move to various distances from the center of the sphere; and that the shortest path between two points, that is to say the path that can be described in the least time, is not a straight line, but the arc of a circle orthogonal to the boundary sphere. To the inhabitant of such an universe, it appears infinite in extent; a circle orthogonal to the boundary appears as a straight line. A little consideration will make it clear that the geometry of this universe is in many respects the same as our elementary geometry, and in some fundamental respects very different; the sum of the angles of a triangle is less than two right angles, and in a plane (for instance, a plane through the center of the sphere) it is possible to draw through a point infinitely many lines not intersecting a given line. The recent advances in physics have indicated that our own universe can be best understood in terms of some such basis as that of Poincaré's fanciful domain.

INVERSION

63. We now investigate the important geometric transformation known as inversion.

Definition. Given a circle c, whose center is O and radius r not zero; if P and P' lie on a line through O, and

$$\overline{OP} \cdot \overline{OP'} = r^2,$$

then P and P' are said to be *inverse* to each other with regard to the circle c. The relation between the points, or the operation of determining either when the other is given, is called *inversion*.

From this definition a number of immediate consequences can be read off.

Theorem. *Every point in the plane except the center O of the circle of inversion has a unique inverse. The relation is reciprocal; that is, if P' is the inverse of P, then P is the inverse of P'. With each point exterior to the circle is associated* *an interior point.   Each point on the circle of inversion is its own inverse, and every self-inverse point is on this circle.*

It is evident from the foregoing, that to any figure in the plane corresponds a second figure, such that corresponding points of the two figures are mutually inverse.   Our problem is to ascertain the mutual relations of two such figures; and especially, those properties of the one figure which remain unchanged in the inverse figure.*   We shall find that every circle or straight line is transformed by an inversion into a circle or straight line, and that the angle of intersection of any two curves is transformed into an equal angle reversed in direction.   Because of these two important invariant relationships, the transformation of a figure by inversion is exceedingly useful in geometric investigation.

64. Special cases and conventions.   Because we have agreed to regard straight lines as one kind of circles, we augment our definition of inversion as follows:

The inverse of a point P with regard to a straight line AB is the reflection of P with regard to AB; in other words, the point P' such that AB is the perpendicular bisector of PP'.

(For let P and P' be inverse with regard to a circle $O\ (r)$; let OPP' cut the circle at C.   We have

$$\overline{OP} \cdot \overline{OP'} = r^2;$$

or $\qquad (r + \overline{CP})\ (r + \overline{CP'}) = r^2.$

Hence $\qquad r\ (\overline{CP} + \overline{CP'}) = -\ \overline{CP} \cdot \overline{CP'}$

$$\overline{CP} + \overline{CP'} = -\ \frac{\overline{CP} \cdot \overline{CP'}}{r}$$

---

* We shall, however, in spite of all temptations, refrain from the use of the formidable word *anallagmatic*, which is Greek for *invariant*, and is used with the meaning "unchanged by inversion."

Now let P and C remain fixed in position, while O is removed to a great distance. The circle approaches as limiting from a straight line perpendicular to PP' at C; and $\overline{CP} + \overline{CP'}$ approaches the limit zero. Hence the above definition; when the circle of inversion is replaced by a straight line, inverse points are equidistant from this line on a perpendicular to it.)

On the other hand, inversion with regard to a null-circle is neither feasible nor useful.

Some geometers find it desirable to define as an inversion the transformation determined by the equation

$$\overline{OP} \cdot \overline{OP'} = -c^2$$

the circle of inversion being imaginary, with radius $c \sqrt{-1}$. Since we are avoiding the use of imaginary numbers, we shall achieve such a transformation as this by an inversion followed by rotation of the figure through 180°, which yields the same result.

Another convention which is associated with the study of inversion we merely mention in passing. In an earlier chapter, we introduced ideal elements at infinity, and perhaps their use has already been justified. Now in the geometry of inversion, we see that there is a single point whose inverse does not exist, namely the center of inversion; and as a point recedes to a great distance in any direction, its inverse approaches the center. In the geometry of inversion, therefore, it is usual to sacrifice the line of points at infinity which is so useful in other fields, and adopt the convention of a single point at infinity, the inverse of the center of the circle of inversion. However, in this book we shall not find it necessary to adopt this point of view. It is true that our points at infinity have no inverses; but we shall presently ascertain the fate, under an inversion, of a set of parallel lines, and we shall assume that only the finite plane is affected by an inversion. When the inverse of a point is mentioned, it is to be understood that this point may not be at infinity nor at the center of inversion.

We proceed to establish simple practical constructions for the inverse of a given point.

65. Theorem. *If a point P is outside the circle of inversion O (r), its inverse P′ is the intersection of OP with the line connecting the points of contact of the tangents to the circle from P.*

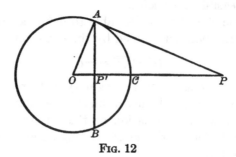

FIG. 12

For let the tangents be PA, PB; and let AB meet OP at P'. Then triangles $AP'O$ and PAO are similar, and

$$\overline{OP'}\cdot\overline{OP} = \overline{OA}^2$$

which establishes P and P' as inverse points.

Theorem. *Conversely, if P is inside the circle, and AB is the chord perpendicular to OP, the tangents to the circle at A and B meet at the inverse point.*

66. Besides these obvious constructions, we have the following, effected with compass alone, and perhaps the most convenient in actual practice. A construction with ruler alone will be given later (**139**).

Theorem. *Let P be any point distant from O more than ½ r. Let the circle P (PO) cut the circle of inversion at X and Y, and let the circles X (XO) and Y (YO) meet at O and P′. Then P′ is the inverse of P.*

The proof, by means of similar triangles, is easy.

Exercise. *Show how to find the inverse of a point P with compasses only, in case the circle P (PO) does not intersect the circle of inversion.* (It will be seen that in this case the construction, while theoretically possible, is not useful in practice.)

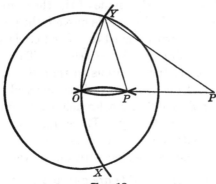

Fig. 13

67. A simple mechanism for effecting inversion is suggested by the following theorem.

Theorem. *If ABCD is a rhombus, and O is equidistant from its opposite vertices A and C, then O, B, D are collinear, and*

$$\overline{OB} \cdot \overline{OD} = \overline{OA}^2 - \overline{AB}^2$$

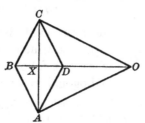

Fig. 14

For $\overline{OB} \cdot \overline{OD} = (\overline{OX} + \overline{XB})(\overline{OX} - \overline{XB})$

$\qquad = \overline{OX}^2 - \overline{XB}^2 = \overline{OX}^2 + \overline{XA}^2 - (XB^2 + XA^2)$

$\qquad = \overline{OA}^2 - \overline{AB}^2$

The Peaucellier cell or *inversor* consists essentially of four equal rods forming a rhombus $ABCD$, and two longer equal rods OA and OC. All the vertices are freely turning joints, and O is attached to the drawing table. Pencils may be attached at B and D. By virtue of the theorem just proved, in whatever manner the linkage is moved, the points B and D will remain inverse with regard to a fixed circle whose center is O and whose radius is $(\overline{OA}^2 - \overline{OB}^2)^{\frac{1}{2}}$. This circle can be drawn by flattening the rhombus so that B and D coincide. Other mechanical inversors have been devised, but this apparatus, invented in 1867 by a French army officer, Peaucellier, is the simplest in theory. The student is urged to construct a model with simple materials, such as strips of cardboard or light wood, connected with brass eyelets or rivets.

68., Theorem. *If P and Q are any points, P' and Q' their inverses, the triangles OPQ and $OQ'P'$ are inversely similar.*

For the angles at O are common to the triangles, and the including sides are proportional.

a. **Corollary.** *Any two pairs of inverse points lie on a circle orthogonal to the circle of inversion.*

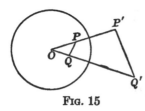

FIG. 15

As before, since
$$\overline{OP}\cdot\overline{OP'} = \overline{OQ}\cdot\overline{OQ'} = r^2$$
the four points lie on a circle, with regard to which the power of O is r^2; whence it follows that this circle is orthogonal to the circle of inversion.

b. **Corollary.** *The distance between two points, and that between the inverse points,* are connected by the equation*

$$\overline{P'Q'} = \overline{QP}\,\frac{r^2}{\overline{OP}\cdot\overline{OQ}}$$

---

* It is to be kept clearly in mind that the points of the line-segments PQ and $P'Q'$, except these points themselves, are not mutually inverse.

For in the similar triangles,

$$\overline{P'Q'} = \overline{QP}\, \frac{\overline{OP'}}{\overline{OQ}} = \overline{QP}\, \frac{r^2}{\overline{OP}} \cdot \frac{1}{\overline{OQ}}$$

c. *For any four points and their inverses,*

$$\frac{\overline{P'Q'} \cdot \overline{R'S'}}{\overline{P'S'} \cdot \overline{R'Q'}} = \frac{\overline{PQ} \cdot \overline{RS}}{\overline{PS} \cdot \overline{RQ}}$$

d. **Exercises.**   Establish the foregoing results in case any of the points, as P and Q, are collinear with the center of inversion. Formulate the analogous theorems for the case that the circle of inversion is a straight line.

69. We are now ready to consider the highly important problem of subjecting to an inversion a circle or a straight line.   Taking the simplest case first:

Theorem.   *A straight line passing through the center of inversion is inverse to itself.*

Corollary.   *Any pair of mutually inverse points divide the diameter of the circle of inversion internally and externally in ratios numerically equal.*

70. Theorem.   *The inverse of any straight line not passing through the center of inversion is a circle through that center; and conversely.*

For let OA be the perpendicular from the center O to the straight line AB.   We have seen that triangles OAB and $OB'A'$ are inversely similar.   Hence as B moves on line AB, triangle $OA'B'$ is a variable right triangle on the fixed hypotenuse OA', and the locus of B' is a circle on OA' as diameter.

71. Theorem.   *The inverse of a circle not passing through the center of inversion is a circle; the center of inversion is a center of similitude of the mutually inverse circles, and any pair of inverse points are antihomologous.*

(It is to be noted, however, that the centers of two mutually inverse circles are not, in general, inverse points.)

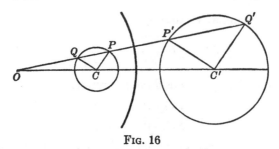

Fig. 16

For let any line through O cut the given circle at P and Q, so that the power of O with regard to the circle is

$$t = \overline{OP} \cdot \overline{OQ}$$

Also, if P' is the inverse of P, and r the radius of inversion,

$$\overline{OP} \cdot \overline{OP'} = r^2$$

Dividing, we have
$$\frac{\overline{OP'}}{\overline{OQ}} = \frac{r^2}{t}$$

We interpret this last equation as showing that when Q describes the given circle, the point P' on the moving line OQ divides that line in the constant ratio r^2/t. Hence (**21, 26**) P' describes a circle simultaneously with Q, and the center of inversion O is a homothetic center for these homologous moving points. It follows at once that on these two circles P and P' are antihomologous points (cf. also **28**). The center O is the external or the internal center of similitude of the two inverse circles, according as t is positive or negative; that is, as O lies outside or within the given circle.

Corollary. *If a circle is orthogonal to the circle of inversion, its points are mutually inverse, and the circle as a whole is unchanged by the inversion.*

Theorem. *If the radius of the given circle is R, and the power of O with regard to it is t (not zero), then the radius R'*

of the inverse circle and the power t' of O with regard to it, are given by

$$R' = R\frac{r^2}{t}, \qquad t' = \frac{r^4}{t}$$

For if the line through the centers of the circles cuts the given circle in P and Q, whose inverses are P' and Q' respectively, then

$$\overline{OP'}\cdot\overline{OQ'} = \frac{r^2}{\overline{OP}}\cdot\frac{r^2}{\overline{OQ}}$$

$$\overline{Q'P'} = \overline{OP'} - \overline{OQ'} = r^2\left(\frac{1}{\overline{OP}} - \frac{1}{\overline{OQ}}\right) = r^2\frac{\overline{PQ}}{\overline{OP}\cdot\overline{OQ}}$$

Corollary. *It is always possible to invert two circles into equal circles.* (See further **129.**)

72. To the Peaucellier inversor, as already described, we may add one more bar, hinged at one end to B and attached at the other end to the table. Thereby B will be constrained to describe a circle, and the inverse point D will simultaneously describe a circle. In particular, if the center of inversion O lies on the circle described by B, then D will traverse a straight line. It is worthy of note that the usual processes of drawing a straight line presuppose the existence of a straight edge already constructed. Until the invention of the Peaucellier inversor there was perhaps no solution of the problem of constructing a straight line *ab initio*, but the inversor furnishes a neat solution of this problem.

73. From the foregoing theorems we derive easily the second fundamental property of inversion.

Theorem. *The tangents to two mutually inverse circles at corresponding points are equally inclined to the line passing through the points and the center of inversion.*

For the mutually inverse points are antihomologous; and we noted in **30** that the tangents to two circles at antihomologous points make equal angles with the line joining the points.

74. Theorem. *The angle between the inverses of two circles equals the angle between the original circles, but is described in the opposite direction.*

For such an angle is the sum or difference of the angles which the respective tangents make with the line joining the mutually inverse points of intersection.

This very important theorem can be established in various ways; as for instance by applying **75** to a secant PQ of a curve, then letting the secant approach tangency. The theorem can thus be generalized to show that any two curves invert into curves intersecting at equal angles; in other words, *the transformation is inversely conformal.*

Theorem. *If two circles are orthogonal, their inverses are also orthogonal. A self-inverse circle or straight line is orthogonal to the circle of inversion.*

75. The following theorem expresses in simple form a highly useful relation involving the angles of mutually inverse figures; indeed, both the fundamental properties of inversion which we have established can be based on it.

Theorem. *If P', Q', R' are respectively the inverses of three points P, Q, R, and O is the center of inversion, then*

$$\angle PQR + \angle P'Q'R' = \angle POR$$

For $\angle PQO = \angle OP'Q' = \angle P'OQ' + \angle OQ'P'$ (**18, 68**)
and similarly
$$\angle OQR = \angle Q'R'O = \angle R'Q'O + \angle Q'OR'$$

Adding,

$$\angle PQO + \angle OQR = \angle R'Q'O + \angle OQ'P' + \angle P'OQ' + \angle Q'OR'$$

Combining and transposing, we have the result as stated.

An alternative statement:

$$\angle P'Q'R' = \angle POR + \angle RQP$$

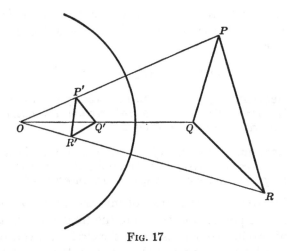

Fig. 17

In either form, the equation gives the relationship between the angles of mutually inverse triangles. It shows, for instance, that two such triangles cannot be inversely similar, and can be directly similar only when each is inscribed in a circle concentric with the circle of inversion. This result follows also from **68 b**.

76. Theorem. *For any four points and their inverses,*

$$\angle PQR + \angle RSP = \angle P'S'R' + \angle R'Q'P'$$

This equation, in connection with **19**, furnishes an immediate proof that the inverse of a circle is a circle.

77. We now formulate a number of theorems which follow easily from the fundamental properties of inversion, and which will be needed frequently hereafter. Other theorems of somewhat lesser applicability will be found in Chapter V.

Theorem. *If a circle passes through two mutually inverse points, it is orthogonal to the circle of inversion, and its points are inverse in pairs.*

For if P and P' are inverse with respect to the circle $O\,(r)$, and if any circle through P and P' is cut by a line through O in Q and Q', then

$$\overline{OP}\cdot\overline{OP'} = \overline{OQ}\cdot\overline{OQ'} = r^2$$

78. Theorem. *If each of two intersecting circles is orthogonal to a third, their points of intersection are mutually inverse with regard to the third circle.*

79. Theorem. *The circle of inversion is coaxal with any two mutually inverse circles. If the circles do not intersect, the limiting points of the coaxal system are inverse points.*

For if a circle cuts the circle of inversion, its inverse cuts it at the same points, and therefore the three are coaxal. If neither of two inverse circles meets the circle of inversion, we draw several circles orthogonal to one of them and to the circle of inversion. Since orthogonal circles invert into orthogonal circles (**74**) the effect of the inversion is that each of these circles is orthogonal to the second circle. Now these auxiliary circles constitute a coaxal system; and the two given circles and the circle of inversion will therefore be members of the conjugate coaxal system. The limiting points of the latter will be the points common to the system of orthogonal circles.

Corollary. *The circle of inversion is coaxal with any pair of inverse points regarded as limiting points. Hence (**60**) the distances from two fixed inverse points to a point moving on the circle of inversion are in a constant ratio.*

This last statement may also be deduced from **68** *b.*

Corollary. *If any circle of a coaxal system is taken as circle of inversion, the circles of the system are interchanged in pairs.*

80. Theorem. *If two points are inverse with regard to a circle, and the whole figure is inverted with regard to another circle, in the resulting figure the points are inverse with regard to the circle.*

That is, if P and Q are inverse with regard to circle c, and the inverses of these with regard to a circle b are P', Q', c', we are to prove that P' and Q' are inverse with regard to c'. We draw any two circles j and k through P and Q; then they are orthogonal to c, and their inverses j' and k' with regard to b are orthogonal to c'. Therefore the intersections P' and Q' of j and k are inverse with regard to c'.

This theorem may be interpreted as showing that "the property of inverseness is invariant under inversion." That is, if two mutually inverse figures, together with their circle of inversion, are subjected to an inversion, the resulting figures are mutually inverse.

81. The following theorems, indicating what changes and simplifications of a figure can be effected by inversion, are established at once. Further questions of the same sort will be discussed later (**129–131**).

a. Any two pairs of points on a circle can be interchanged by an inversion, provided their connectors intersect outside the circle; that is, provided they do not separate each other.

The same theorem in the latter form holds when the four points are on a line.

b. Two or more circles through a point may be inverted into straight lines, taking the common point as center of inversion; and conversely, the straight lines of a plane can be inverted into the totality of circles through a point.

c. Two or more circles, tangent at a point, will be inverted into parallel lines if the center of inversion is at the point of tangency; and conversely.

d. Two circles that do not intersect will be inverted into con-

centric circles, if either limiting point of their coaxal system is taken as center of inversion.

Summarizing the foregoing, two conjugate coaxal systems of the first and second type respectively will be transformed, by an inversion whose center is one of the fixed points of the systems, into a set of concentric circles and a set of lines radiating from the common center. This center is the inverse of the second fixed point. Thus the properties of coaxal systems may be derived by inversion from those of the special limiting coaxal systems of types IV and V (**54**).

Two conjugate coaxal systems of the third type, each consisting of circles tangent to a common radical axis at a point, may be transformed by inversion into two mutually perpendicular sets of parallel lines. In **55**, this figure was noted as one of the special types of conjugate coaxal systems.

82. Theorem. *The inverse of the center of a given circle is the same as the inverse of the center of inversion with regard to the circle inverse to the given circle.*

That is, if circles k and k' are inverse with regard to a circle c, whose center is O; and if A is the center of k, then A', the inverse of A with regard to c, is the same as the inverse of O with regard to k'.

For let us draw a set of straight lines through A; being orthogonal to k, they invert (with regard to c) into a set of circles through O and A' and orthogonal to k'; whence O and A' are inverse with regard to k'.

Corollary. *To construct the inverse of a given circle by locating its center, find the inverse of the center of inversion with regard to it, and then invert this point with regard to the circle of inversion.*

Exercise. *Modify the foregoing statements to apply to the case that the circle of inversion becomes a straight line.* We recognize that two figures inverse with regard to a straight

line are symmetrically congruent; and it may be possible
to prove theorems concerning two mutually inverse figures
by transforming their circle of inversion to a straight line.

Exercise. *If two circles are orthogonal, the inverse of either
center with regard to the other circle is the mid-point of the
common chord.*

Exercise. Complete the proofs of all propositions in the
latter half of this chapter, viz: **63, 65, 66, 69, 71** (corollaries), **76, 78, 79** (corollaries), **81**.

CHAPTER IV

TRIANGLES AND POLYGONS

83. The present chapter is an assemblage of various theorems concerning triangles, quadrilaterals, and other polygons, all of which can be established at once on the basis of the familiar elementary geometry and the results that have thus far been obtained. Considerable portions of this chapter, as well as most of Chapters V and VI, may be omitted without destroying the sequence. The reader who is desirous of proceeding at once to the general theory of the triangle, without delaying for material which, however interesting, is irrelevant to the main purpose, will read sections **84–92, 95–101, 104** a, and then pass on to Chapter V.

84. We have first a theorem which is highly useful in the study of the triangle, because of the various forms which it assumes and the different corollaries which can be at once deduced from it.

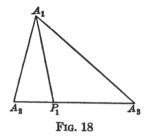

FIG. 18

Theorem. *If P_1 is any point except A_3 of the side A_2A_3 of a triangle $A_1A_2A_3$, then*

$$\frac{\overline{P_1A_2}}{\overline{P_1A_3}} = \frac{\overline{A_1A_2}\,\sin\angle P_1A_1A_2}{\overline{A_1A_3}\,\sin\angle P_1A_1A_3}$$

For by the law of sines (**15** b)

$$\frac{\overline{P_1A_2}}{\overline{A_1A_2}} = \frac{\sin\angle P_1A_1A_2}{\sin\angle A_1P_1A_2} \quad \text{and} \quad \frac{\overline{A_1A_3}}{\overline{P_1A_3}} = \frac{\sin\angle A_1P_1A_3}{\sin\angle P_1A_1A_3}$$

Since also $\quad \sin\angle A_1P_1A_2 = \sin\angle A_1P_1A_3$

the theorem follows at once by combining these equations.

Corollary. *If A_1P_1 and A_1Q_1 are so drawn that angles $A_2A_1P_1$ and $Q_1A_1A_3$ are equal, then*

$$\frac{\overline{P_1A_2}\cdot\overline{Q_1A_2}}{\overline{P_1A_3}\cdot\overline{Q_1A_3}} = \left(\frac{\overline{A_1A_2}}{\overline{A_1A_3}}\right)^2$$

85. Theorem. *If P_1, P_2, P_3 are on the sides A_2A_3, A_3A_1, A_1A_2, respectively, of triangle $A_1A_2A_3$,*

$$\frac{\overline{P_1A_2}\cdot\overline{P_2A_3}\cdot\overline{P_3A_1}}{\overline{P_1A_3}\cdot\overline{P_2A_1}\cdot\overline{P_3A_2}} = \frac{\sin\angle P_1A_1A_2\cdot\sin\angle P_2A_2A_3\cdot\sin\angle P_3A_3A_1}{\sin\angle P_1A_1A_3\cdot\sin\angle P_2A_2A_1\cdot\sin\angle P_3A_3A_2}$$

That is to say, the product of the ratios in which the sides of the triangle are divided equals the corresponding product for the sines of the angles subtended at the opposite vertices.

86. Theorem. *If P_1, Q_2 are any points on A_2A_3, the "double ratio" composed of the ratio of the ratios in which they divide A_2A_3 equals the corresponding double ratio for the sines of the angles subtended at A_1;*

$$\frac{\overline{P_1A_2}}{\overline{P_1A_3}} : \frac{\overline{Q_1A_2}}{\overline{Q_1A_3}} = \frac{\sin\angle P_1A_1A_2}{\sin\angle P_1A_1A_3} : \frac{\sin\angle Q_1A_1A_2}{\sin\angle Q_1A_1A_3}$$

Theorem. *If three lines, concurrent at O, are cut by a transversal at A, B, C, and by another at A', B', C', then*

$$\frac{\overline{AB}}{\overline{AC}} : \frac{\overline{A'B'}}{\overline{A'C'}} = \frac{\overline{OB}}{\overline{OC}} : \frac{\overline{OB'}}{\overline{OC'}}$$

87. Theorem. *If four concurrent lines meet one transversal at A_1, A_2, A_3, A_4, and another at B_1, B_2, B_3, B_4, then*

$$\frac{\overline{A_1A_3}}{\overline{A_1A_4}} : \frac{\overline{A_2A_3}}{\overline{A_2A_4}} = \frac{\overline{B_1B_3}}{\overline{B_1B_4}} : \frac{\overline{B_2B_3}}{\overline{B_2B_4}}$$

This fundamental theorem of projective geometry is an immediate consequence of **86**.

Definition. The double ratio, or ratio of the ratios of the distances of two pairs of points on a line, as discussed above,

is called the *cross-ratio* or *anharmonic ratio* of the four points.
In the particular case that its value is -1, we see that each
pair of points separate the other pair internally and externally in the same ratio; they are said to separate one
another *harmonically*, or to constitute a *harmonic set* of
points.   Again, the cross-ratio or anharmonic ratio of four
concurrent lines is the double ratio, as given in **86**, of the
sines of the angles between them.   It is equal to the cross-
ratio of the points in which the lines are cut by any transversal.   Four concurrent lines form a harmonic set, when
their cross-ratio is -1.

88. Projective geometry.   Let us consider two planes, generally not parallel, and a point O which is not in either plane.
From the points of one plane, lines may be drawn through O,
cutting the second plane.   Thus any figure in the first plane
is "projected" into a figure in the second plane.   Projective
geometry may be defined as the study of those properties
of a figure which are unaltered by such projections, and such
properties are called *projective*.   For instance, we see that in
general projective figures are not similar; properties involving
ratios and angles are not projective properties of a figure.
On the other hand, any relationship involving concurrence of
lines and collinearity of points is projective.

The theorems of the last three sections have established the
fact that the cross-ratio of four points on a line is pro-
jectively invariant, and is equal to the corresponding double
ratio for the sines of the angles between the projecting rays.
This is fundamental in projective geometry.

We shall not explore the domain of projective geometry,
except occasionally to survey it from the frontier which it
shares with our field.   We owe the concept of the line at
infinity to projective geometry; but for the most part, since
the familiar relations of distances, angles and ratios are not
invariant in that study, we have little in common with it.
Occasionally, as in Chapter XIII, we shall note theorems

which are projective in nature; and the concept of a harmonic set of points or lines will sometimes be useful to us.

QUADRANGLES AND QUADRILATERALS

89. Let us establish a few definitions and conventions concerning polygons, especially quadrilaterals.

Definitions. A *simple quadrangle* or *quadrilateral* is a closed polygon with four vertices and four sides. The connectors of the pairs of vertices not already connected are the two *diagonals*.

A *complete quadrangle* is the figure determined by four points in general position, and their six connectors. Two connectors not having in common any one of the given points are called *opposite*; there are three such pairs, and the points of intersection of opposite connectors are called *diagonal points*.

A *complete quadrilateral* is the figure composed of four lines in general position, and their six points of intersection. There are three pairs of *opposite points*, whose connectors are the *three diagonals*.

Theorems concerning the complete quadrilateral and quadrangle will appear from time to time. One famous theorem of the quadrilateral follows below, after the proof of a lemma on which its proof is based.

90. Theorem. (Euclid, I, 43.) *If through any point of a diagonal of a parallelogram we draw lines parallel to the sides, the parallelograms not containing segments of that diagonal are equal in area, and conversely.*

That is, if through X, a point of the diagonal AC of parallelogram $ABCD$, we draw lines parallel to the sides of the parallelogram, the areas BX and DX are equal. For the diagonal AC bisects each of the areas AC, AX, XC; whence the result by subtracting equals from equals.

Conversely, if two lines parallel to AB and BC meet at X, making areas BX and XD equal, then X lies on AC.

This old theorem, used by Euclid as a first step in the comparison of areas, and leading up to his proof of the theorem of Pythagoras, has been generally dropped from the modern geometries. It can be made useful occasionally, as for instance in problems of construction relating to equal areas. It yields easily a proof of the following well-known theorem.

91. Theorem. *The mid-points of the diagonals of a complete quadrilateral are collinear.*

Denoting the four lines of the complete quadrilateral by ABC, $AB'C'$, $A'BC'$, $A'B'C$, the three diagonals are AA',

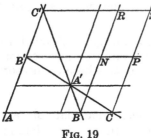

FIG. 19

BB', CC'. We draw construction lines parallel to two of the four lines, and letter as in the figure. Then applying the theorem of **90**:

$$\text{area } AA' = \text{area } A'R$$
$$\text{area } AA' = \text{area } A'P$$
$$\text{Hence} \quad \text{area } A'R = \text{area } A'P,$$

and therefore A' is on the diagonal NS. Thus the mid-points of AA', AN, and AS are collinear. But AS and CC' bisect each other, as do also AN and BB'. We have therefore proved that the mid-points of AA', BB', CC' are collinear.

Later (**268**) we shall establish this theorem again, showing further that the three circles on AA', BB', and CC' are coaxal, and proving other theorems about the four triangles formed by the four lines.

THE THEOREM OF PTOLEMY

92. Theorem. *If a quadrangle is inscribed in a circle, the product of the diagonals equals the sum of the products of the* *opposite sides. Conversely, if the sum of two opposite connectors of four points equals the sum of the products of the other pairs, the four points lie on a circle.*

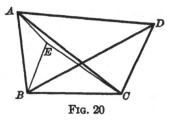

FIG. 20

Both theorems can be proved almost simultaneously. Let A, B, C, D be any four points, whereof B, C, D are not collinear. On AB construct the triangle ABE, directly similar to DBC. Thus,

$$\overline{BD}\cdot\overline{AE} = \overline{AB}\cdot\overline{DC}$$

Further, $\quad \dfrac{\overline{BD}}{\overline{BC}} = \dfrac{\overline{BA}}{\overline{BE}} \qquad$ and $\angle ABD = \angle EBC$

so that triangles ABD and EBC are also similar, and

$$\overline{BD}\cdot\overline{EC} = \overline{BC}\cdot\overline{AD}$$

Adding, $\qquad \overline{BD}\,(\overline{AE} + \overline{EC}) = \overline{AB}\cdot\overline{CD} + \overline{AD}\cdot\overline{BC}$

Now, E will lie on AC if and only if

$$\angle BAE = \angle BAC = \angle BDC$$

which is true if and only if A, B, C, D are concyclic. That is, if the four points are concyclic,

$$\overline{AE} + \overline{EC} = \overline{AC}$$

and in all other cases,

$$\overline{AE} + \overline{EC} > \overline{AC}$$

Accordingly, the sum of the products $\overline{AB}\cdot\overline{CD} + \overline{AD}\cdot\overline{BC}$ will equal or exceed $\overline{AC}\cdot\overline{BD}$ according as A, B, C, D are or are not concyclic.

The foregoing proof is inapplicable if and only if the four

points are collinear; and we proved in **3** that the equation holds for any four collinear points.

Second proof: as before, let A, B, C, D be any four points; taking D as center of inversion, let the inverses of A, B, C be respectively A', B', C'. Now A', B', C' will be collinear if and only if A, B, C, D are on a circle. If this condition is satisfied,

$$\overline{A'B'} + \overline{B'C'} + \overline{C'A'} = 0$$

But we recall **68** b and make the substitution for each of these lengths:

$$\overline{AB}\, \frac{r^2}{\overline{DA}\cdot\overline{DB}} \pm \overline{BC}\, \frac{r^2}{\overline{DB}\cdot\overline{DC}} \pm \overline{CA}\, \frac{r^2}{\overline{DC}\cdot\overline{DA}} = 0$$

Clearing of fractions,

$$\overline{AB}\cdot\overline{CD} \pm \overline{AC}\cdot\overline{DB} \pm \overline{AD}\cdot\overline{BC} = 0$$

an equation which is true if and only if the four points are on a circle.

93. Numerous theorems of geometry and trigonometry can be deduced as consequences of the theorem of Ptolemy.

a. Noting that $2R \sin \phi$ represents the length of a chord in a circle of radius R, whose central angle is 2ϕ, the addition formulas of trigonometry may be derived at once from the theorem of Ptolemy:

$$sin\ (a + b) = sin\ a\ cos\ b + cos\ a\ sin\ b,\ etc.$$

b. If ABC is an equilateral triangle, and P lies on the arc BC of the circle through A, B, C, then

$$\overline{PC} = \overline{PA} + \overline{PB}$$

c. If D is on the arc BC of the circumscribed circle of an isosceles triangle ABC, with $\overline{AB} = \overline{AC}$, then

$$\frac{\overline{PA}}{\overline{PB} + \overline{PC}} = \frac{\overline{AB}}{\overline{BC}},\ a\ constant\ ratio.$$

d. If P is on the arc AB of the circle circumscribed about a square ABCD, then

$$\frac{\overline{PA} + \overline{PC}}{\overline{PB} + \overline{PD}} = \frac{\overline{PD}}{\overline{PC}}$$

e. If P lies on the arc AB of the circle circumscribed about a regular hexagon ABCDEF, then

$$\overline{PD} + \overline{PE} = \overline{PA} + \overline{PB} + \overline{PC} + \overline{PF}$$

f. If P lies on the arc AB of the circle circumscribed to a regular pentagon ABCDE, then

$$\overline{PC} + \overline{PE} = \overline{PA} + \overline{PB} + \overline{PD}$$

Such theorems can be spun on indefinitely. We close the group with a somewhat more elaborate theorem, involving the sides and principal diagonals of a hexagon inscribed in a circle.

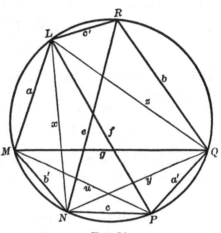

FIG. 21

g. **Theorem.**[*] *Let the opposite sides of a convex hexagon inscribed in a circle be a, a'; b, b'; c, c'; and let the diagonals be e, f, g (so chosen that a, a', and e have no common vertex, nor have b, b', f), then*

$$efg = aa'e + bb'f + cc'g + abc + a'b'c'$$

Let the hexagon be *LMNPQR*, with $\overline{LM}$, $\overline{MN}$, etc., in order, denoted by *a, b', c, a', b, c'*; then $\overline{NR}$, $\overline{LP}$ $\overline{MQ}$, are respectively e, f, g. Designate $\overline{LN}, \overline{NQ}, \overline{QL}, \overline{MP}$ by x, y, z, u respectively. Then

$$b'f + ac = ux \qquad \text{and} \qquad cg + a'b' = uy$$

Multiplying by b and c' respectively and adding, we get

$$cc'g + bb'f + abc + a'b'c' = u\,(bx + c'y)$$
$$= uez = euz = e\,(fg - aa')$$

which leads to the desired formula. This theorem may be regarded as an extension of the theorem of Ptolemy to the hexagon.

94. Theorem. *For any four points A, B, C, D*

$$\overline{AC}^2 \cdot \overline{BD}^2 = \overline{AB}^2 \cdot \overline{CD}^2 + \overline{AD}^2 \cdot \overline{BC}^2$$
$$- 2\,\overline{AB} \cdot \overline{BC} \cdot \overline{CD} \cdot \overline{DA}\ \cos\,(\angle ABC + \angle CDA)$$

This generalization of the theorem of Ptolemy is proved by the same method used in the inversion proof of that theorem. If we state the law of cosines for the three points A', B', C',

$$\overline{A'C'}^2 = \overline{A'B'}^2 + \overline{B'C'}^2 - 2\,\overline{A'B'} \cdot \overline{B'C'} \cdot \cos \angle C'B'A',$$

then subject the figure to an inversion whose center is D, and replace each of the lengths by its equivalent, we have at once the formula as given.

95. There are some interesting questions associated with the ratios of the distances of a point from three fixed points. It is a natural procedure to fix the position of a point by giving its distances, or the ratios of its distances, from the vertices of a fixed triangle. This determination, however, is not unique.

Theorem. *There are at most two points in the plane whose distances from three fixed points are proportional to given numbers.*

For let A_1, A_2, A_3 be the given points, and let a point P be sought, such that

$$\overline{PA}_1 : \overline{PA}_2 : \overline{PA}_3 = p_1 : p_2 : p_3$$

where p_1, p_2, p_3 are given numbers.

The locus of a point P, moving so that $\overline{PA}_2 / \overline{PA}_3 = p_2/p_3$, we have seen in **58** to be a circle of the coaxal system whose limiting points are A_2 and A_3. Similarly, the locus of a point P for which $\overline{PA}_3 / \overline{PA}_1 = p_3/p_1$ is a circle coaxal with A_3 and A_1; and so for $\overline{PA}_1$ and $\overline{PA}_2$. But all three of these circles are orthogonal to the circle $A_1A_2A_3$, and there are three cases:

If no two of the three circles intersect, there is no point satisfying the given conditions. If two of the circles intersect, the third must pass through the points of intersection. These two points, which are the solutions of the problem, are therefore mutually inverse with regard to the circumscribed circle $A_1A_2A_3$. Similarly, if two of the circles are tangent, the third is tangent to them at the same point, and this point is the only solution of the problem and lies on the circle $A_1A_2A_3$.

The inquiry next suggests itself, for what values of the ratios $p_1 : p_2 : p_3$ do such points exist?

Theorem. *There exist two points P, P', the ratios of whose distances from three given points A_1, A_2, A_3 are proportional to three given numbers p_1, p_2, p_3, if and only if the three products $p_1 \cdot \overline{A_2A_3}$, $p_2 \cdot \overline{A_3A_1}$, $p_3 \cdot \overline{A_1A_2}$ have the property that the sum of any two of them is less than the third; that is, that these three products are the sides of a possible triangle. If the sum of two of these products equals the third, then there is one such point, situated on the circle $A_1A_2A_3$; and conversely.*

That the inequality holds when the points P, P' exist, is a consequence of Ptolemy's theorem, as also the fact that when there is a single point on the circumcircle the equation is satisfied. The converse theorem, that when the inequality is given we can establish the existence of the points P, P' is not easy at this stage. Using **58** b, c, in conjunction with the proof of the foregoing theorems, we can with considerable labor show that the condition that two of the circles intersect is precisely as stated in the present theorem. We shall, however, a little later be able to establish these results in a simple and elegant manner (**205, 206**).

96. From one of the less familiar theorems of elementary geometry we can derive a number of theorems.

Theorem. *The sum of the squares of two sides of a triangle equals half the square of the third side, plus twice the square of the median on the third side,*

$$a_2^2 + a_3^2 = \tfrac{1}{2} a_1^2 + 2 m_1^2$$

This theorem, which is given in the school textbooks, is proved by applying **14** c to the triangles $A_1A_2O_1$ and $A_1A_3O_1$ (figure 1) and adding the resulting equations. The theorem suggests the following immediate consequences.

a. The length of any median is given by the formula:
$$m_1^2 = \tfrac{1}{4}(2 a_2^2 + 2 a_3^2 - a_1^2)$$

b. $$m_1^2 + m_2^2 + m_3^2 = \tfrac{3}{4}(a_1^2 + a_2^2 + a_3^2)$$

c. $$\overline{MA_1}^2 + \overline{MA_2}^2 + \overline{MA_3}^2 = \tfrac{1}{3}(a_1^2 + a_2^2 + a_3^2)$$

97. Theorem. *The locus of a point which moves so that the sum of the squares of its distances from two fixed points is constant is a circle whose center is midway between the fixed points.*

For in **96** a, if $a_2^2 + a_3^2$ is constant, as well as a_1, then m_1 is constant, and the point moves on a circle of radius m_1.

Theorem. *In a simple quadrangle, the sum of the squares of the four sides is equal to the sum of the squares of the diagonals, plus four times the square of the line joining the midpoints of the diagonals.*

For if $ABCD$ is a quadrangle, and E and F are the mid-points of the diagonals AC, BD, then AF is a median of triangle ABD, and CF of triangle BCD. Hence

$$\overline{AB}^2 + \overline{AD}^2 = \tfrac{1}{2}\overline{BD}^2 + 2\,\overline{AF}^2$$

$$\overline{BC}^2 + \overline{CD}^2 = \tfrac{1}{2}\overline{BD}^2 + 2\,\overline{CF}^2$$

As we add these two equations, we note that EF is a median of triangle ACF, so that

$$\overline{AF}^2 + \overline{CF}^2 = \tfrac{1}{2}\overline{AC}^2 + 2\,\overline{EF}^2$$

whence the result:

$$\overline{AB}^2 + \overline{BC}^2 + \overline{CD}^2 + \overline{DA}^2 = \overline{AC}^2 + \overline{BD}^2 + 4\,\overline{EF}^2$$

Corollary. *In a parallelogram, the sum of the squares of the sides equals the sum of the squares of the diagonals; and conversely, if a quadrilateral has this property, it is a parallelogram.*

Theorem. *In a complete quadrangle, the sum of the squares of any two opposite connectors, plus four times the square of the connector of their mid-points, equals the sum of the squares of the other four connectors.*

Theorem. *The sum of the squares of the six connectors of a complete quadrangle is equal to four times the sum of the squares of the three connectors of mid-points of opposite sides.*

98. We continue with similar elaborations of some other simple leitmotifs.

Theorem. *The difference between the squares of two sides of a triangle equals twice the product of the third side by the projection of the median on that side,*

$$a_2{}^2 - a_3{}^2 = 2\,a_1 \cdot \overline{H_1 O_1}$$

Corollary. *With due regard for signs,*

$$a_1 \cdot \overline{O_1 H_1} + a_2 \cdot \overline{O_2 H_2} + a_3 \cdot \overline{O_3 H_3} = 0$$

Corollary. *If ϕ_1 represents the angle which the median forms with the side on which it rests,*

$$\cot \phi_1 = \frac{a_2^2 - a_3^2}{4\,\Delta}$$

whence $\cot \phi_1 + \cot \phi_2 + \cot \phi_3 = 0$

99. Theorem. *The square of the bisector of an angle of a triangle equals the product of the adjacent sides, minus the product of the segments in which the bisector divides the opposite sides.*

Corollary. *The length of the bisector of angle A_1 is given by*

$$t_1^2 = a_2 a_3 \left(1 - \frac{a_1^2}{(a_2 + a_3)^2} \right)$$

100. Some of the foregoing theorems are special cases of a general theorem due to Apollonius (third century B.C.).

Theorem. *If P_1 is a point dividing the side $A_2 A_3$ of a triangle $A_1 A_2 A_3$ in the ratio $- m/n$, then*

$$ma_2^2 + na_3^2 = (m + n)\,\overline{A_1 P_1}^2 + m\,\overline{P_1 A_3}^2 + n\,\overline{P_1 A_2}^2$$

For $a_2^2 = \overline{A_1 P_1}^2 + \overline{A_3 P_1}^2 - 2\,\overline{P_1 A_3} \cdot \overline{A_1 P_1} \cos \angle A_1 P_1 A_3$

$a_3^2 = \overline{A_1 P_1}^2 + \overline{P_1 A_2}^2 - 2\,\overline{P_1 A_2} \cdot \overline{A_1 P_1} \cos \angle A_1 P_1 A_2$

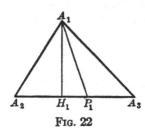

FIG. 22

Multiplying by m and n respectively and adding, we eliminate the terms involving the cosine, and obtain the result as stated.

Replacing $\overline{P_1 A_2}$ and $\overline{P_1 A_3}$ by their values $\dfrac{m}{m + n}\,\overline{A_2 A_3}$ and $\dfrac{n}{m + n}\,\overline{A_2 A_3}$,

and transposing, we have another form:

$$\overline{A_1 P_1}^2 = \frac{m}{m + n}\,a_2^2 + \frac{n}{m + n}\,a_3^2 - \frac{m}{m + n} \cdot \frac{n}{m + n}\,a_1^2$$

101. Theorem. *The product of two sides of a triangle is equal to the product of the altitude on the third side by the diameter of the circumscribed circle.*

The theorem, given in all geometries, is easily proved by similar triangles; it lends itself to a surprising number of variations.

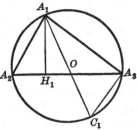

FIG. 23

a. **Corollary.** $\quad 2R = \dfrac{a_2 a_3}{h_1}$;

again, $R = \dfrac{a_1 a_2 a_3}{2\,a_1 h_1} = \dfrac{a_1 a_2 a_3}{4\Delta}$ (cf. **15** *d*).

b. **Theorem.** *If two chords are drawn from a point P on a circle, their product equals the diameter of the circle multiplied by the perpendicular from P on the line joining their extremities.*

c. **Theorem.** *The distance from a point on a circle to a fixed chord, multiplied by the diameter of the circle, equals the product of the distances from the point to the ends of the fixed chord.*

d. **Theorem.** *The distance between two points on a circle is a mean proportional between the diameter of the circle and the perpendicular from either point on the tangent to the circle at the other point.*

This is a limiting case of the previous theorem, or may be proved directly.

e. **Theorem.** *Let the six connectors of four points on a circle be drawn. The products of the perpendicular distances from any other point on the circle to the three pairs of opposite connectors are equal.*

For such a product equals the product of the distances from the point to the four given points, divided by the square of the diameter. This theorem, in turn, suggests some further generalizations.

f. **Theorem.** *Let $A_1A_2A_3 \ldots A_n$ be an even number of points on a circle. Denote by d_{12} the length of the perpendicular from a fixed point P of the circle to the connector A_1A_2, and so on. Then if we form a product of $\frac{1}{2} n$ of the d's, so that each subscript appears once and only once, all such products have the same value.*

As a further extension, we may make some of the A's coincide, whereupon some of the d's become perpendiculars to the tangents to the circle at certain vertices of the remaining polygon. We are thus led to a complicated theorem concerning the perpendiculars to the sides of a polygon and to the tangents at as many of its vertices as we please. Without making any effort to state the theorem in its generality, we note the following case of special interest:

g. **Theorem.** *If a polygon is inscribed in a circle, and a second polygon is circumscribed by drawing tangents to the circle at the vertices of the first, the product of the perpendiculars on the sides of the first, from a point of the circle, equals the product of the perpendiculars from the same point to the sides of the second.*

102. Theorem. *The algebraic sum of the perpendiculars from any point to the sides of a regular polygon of n sides is constant, and equal to n times the apothem.*

(The signs are so attached to these perpendiculars that for a point inside the polygon they are all positive.)

Let a denote each of the equal sides, h the apothem; and let $h_1, h_2, \ldots h_n$ be the perpendiculars from a point P to the sides. Then the area of the polygon is given by

$$\tfrac{1}{2} nha = \tfrac{1}{2} (ah_1 + ah_2 + \ldots + ah_n)$$

so that at once $nh = h_1 + h_2 + \ldots + h_n$

a. For example, the sum of the perpendiculars from any point to the sides of an equilateral triangle equals the altitude; of a square, twice the side of the square; and so on.

b. **Theorem.** *The sum of the perpendiculars from the vertices of a regular polygon to any line tangent to the circumscribed circle is equal to n times the radius.*

This follows from the preceding theorem, by virtue of the fact that the perpendiculars on two tangents to a circle, each from the point of tangency of the other, are equal.

c. **Theorem.** *The algebraic sum of the perpendiculars to any line from the vertices of a regular polygon is equal to n times the distance from the center to the line.*

For if we draw a tangent to the circle parallel to the given line, we may apply the previous theorem.

d. **Theorem.** *The sum of the squares of the distances from any point on a circle to the vertices of a regular polygon of n sides inscribed in the circle is constant, and equal to $2n R^2$.*

For let $A_1 A_2 .. A_n$ be the regular polygon, and P any point on the circumscribed circle. Let the tangent be drawn at P, and denote the perpendicular to this tangent from A_1 by p_1. Then by **101** *d*,

$$\overline{PA_1}^2 = 2 R p_1, \text{ etc.}$$

But as above, the sum of the p's equals nR.

Hence $\quad \overline{PA_1}^2 + \overline{PA_2}^2 + \ldots + \overline{PA_n}^2 = 2 R(p_1 + p_2 + \ldots + p_n)$
$$= 2 n R^2$$

Corollary. *The sum of the squares of the distances from a point on a circle to the mid-points of the sides of a regular inscribed polygon of n sides, a being the length of the side and R the radius of the circle, is*

$$2 n R^2 - \tfrac{1}{4} n a^2$$

This follows at once by application of **96.**

Corollary. *The sum of the squares of all the connectors of the vertices of a regular polygon of n sides inscribed in a circle is $n^2 R^2$.*

For if we apply d, placing P at each of the n vertices in turn, and add the resulting equations, we shall have counted each connector twice, and we shall have the sum $2\,n^2R^2$.

103. *a.* **Theorem.**   *The distance between two points, and the distance between their inverses, are proportional to the perpendiculars from the center of inversion to the two lines.*

For the two pairs of inverse points form with the center of inversion two similar triangles, in which the perpendiculars in question are homologous altitudes.

b. **Theorem.**   *If the lengths of the sides of a polygon inscribed in a circle are a_1, a_2, ... a_n, and if the perpendiculars on these lines from any point P of the circle are p_1, p_2, ... p_n, then with signs properly chosen*

$$\frac{a_1}{p_1} + \frac{a_2}{p_2} + \ldots + \frac{a_n}{p_n} = 0$$

For an inversion with regard to P carries the vertices of the polygon into collinear points.   In the inverted figure all the p's are equal, and

$$a_1' + a_2' + \ldots a_n' = 0$$

c. **Theorem.**   *If p_1, p_2, p_3 are the perpendiculars to the sides of a triangle from any point P, and h_1, h_2, h_3 the altitudes, then:*

$$\frac{p_1}{h_1} + \frac{p_2}{h_2} + \frac{p_3}{h_3} = 1$$

For
$$\frac{p_1}{h_1} = \frac{\text{area } A_2A_3P}{\text{area } A_1A_2A_3}, \text{ etc.}$$

104. We continue with a number of unrelated theorems, dealing mainly with circles.   To a considerable extent the proofs are left to the reader.

a. **Theorem.** *If three equal circles pass through a point, the circle through their other three intersections is equal to them.*

This theorem and the corollaries which follow are established by means of the radii of the circles drawn to the various points of intersection, which form several parallelograms. Hence:

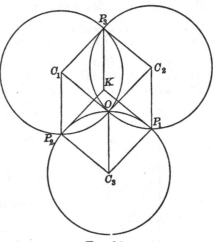

FIG. 24

Corollary. *In this figure, the four points of intersection of the circles form a figure congruent to that of the four centers, with sides parallel and extended in reverse directions.*

Corollary. *In either of these congruent figures, the line joining any two points is perpendicular to that joining the other two.* (Cf. **260**.)

Corollary. *By an inversion, any three lines tangent to the circle of inversion and the circle circumscribed to their triangle, are transformed into equal circles.*

b. **Theorem.** *Let two circles with centers at O and O' intersect at P and Q. Let AB be a diameter of the first, and let AP and BP meet the second circle at A' and B' respectively. Then $A'B'$ is a diameter of the second circle; the angle between AB and $A'B'$ equals the angle of intersection of the circles, namely OPO'; the point of intersection X of AB and AB' lies on the circle OQO'.*

c. **Theorem.** *Let the four common tangents to two mutually external circles be drawn. The points of contact of the direct tangents, those of the transverse tangents, and the intersections of direct with transverse tangents, lie respectively* *on circles whose common center is the point midway between the centers of the given circles.*

d. Theorem.   *Through one of the common points of two circles a variable line is drawn.   Its length between the variable points of intersection with the circles is proportional to the sine of the angle which it makes with the common chord.*

Corollary.   *The longest such line is perpendicular to the common chord; and lines equally inclined to the latter are equal.*

e. Theorem.   *If a variable tangent to a circle meets two fixed parallel tangents AB and BQ at P and Q respectively, then PQ subtends a right angle at the center of the circle, and the radius is a mean proportional between AP and BQ.*

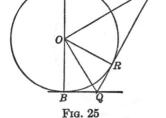

FIG. 25

This result can perhaps be utilized as a convenient construction for proportional lines.   Further:

Corollary.   *The segments cut on parallel tangents to a circle by a variable tangent are inversely proportional.*

Similar theorems of the sort are not unusual.

f. Theorem.   *Let ABC be an isosceles triangle, D the midpoint of the base BC; let P and Q be chosen on AC and AB respectively, so that $\overline{GB} \cdot \overline{CP}$ equals $\overline{BD}^2$; then PQ is tangent to a fixed circle whose center is D and which touches AB and AC.*

g. The following, which is not so simple, is interesting on account of the famous names associated with it.   It is attributed to Fermat, and the earliest proofs which we have are due to Euler and to Simson.   The proof given is that of Fuortes.*

Theorem.   *On one side of a segment AB, a semicircle is drawn.   On the other, a rectangle ABDC is constructed, with*

* *Giornale di matematiche,* 1869; see further Simon, *l.c.,* p. 88. The intrinsic importance of this theorem is not evident.

altitude AC equal to the side of the square inscribed in the circle, viz. $\overline{AB}/\sqrt{2}$. If now from any point P of the semicircle, PD and PC are drawn, cutting AB at E and F respectively, then

$$\overline{AE}^2 + \overline{BF}^2 = \overline{AB}^2$$

Proof: If PA and PB are extended to meet CD at M, N respectively, we have

$$\overline{CN}^2 = \overline{CD}^2 + \overline{DN}^2 + 2\,\overline{CD}\cdot\overline{DN}$$

But triangles AMC and NBD are similar, hence

$$\overline{MC}\cdot\overline{DN} = \overline{AC}\cdot\overline{BD} = \overline{AC}^2 = \tfrac{1}{2}\,\overline{AB}^2$$
$$\overline{CD}^2 = 2\,\overline{AC}^2 = 2\,\overline{MC}\cdot\overline{DN}$$

Adding $\overline{MD}^2$,

$$\overline{MD}^2 + \overline{CN}^2 = \overline{MD}^2 + \overline{DN}^2 + 2\,\overline{CD}\cdot\overline{DN} + 2\,\overline{MC}\cdot\overline{DN}$$
$$= \overline{MD}^2 + \overline{DN}^2 + 2\,\overline{MD}\cdot\overline{DN} = \overline{MN}^2$$

But $\overline{AF}, \overline{FE}, \overline{EB}$ are proportional to $\overline{MC}, \overline{CD}, \overline{DN}$, hence the desired result.

h. **Theorem.**[*] *Let ABC be an isosceles triangle, with $\overline{AB} = \overline{AC}$. With any points P and Q on AB as centers, draw circles p, q passing through B; and with centers R, S lying on AC, draw circles r, s passing through C. Let p and r meet at X and Y, q and s at Z and W.*
If PR and QS meet at a point T, then T is the center of a circle passing through X, Y, Z, W; if PR and QS are parallel, X, Y, Z, W lie on a line perpendicular to them.

This theorem, apparently difficult, is easily proved by means of the power relation. For if the perpendiculars to AB and AC at B and C meet at D, then DB and DC are equal tangents to the four circles, and these four circles are

* Affolter, *Math. Annalen*, vi. 1873, p. 596.

orthogonal to the circle D (DB). Hence any two pairs of their intersections lie on a circle orthogonal to this circle; and it is easy to see that T is the center of such a circle.

Interesting corollaries and special cases may be noted.

105. Theorem. *Through P, the mid-point of a chord l of a circle, let any chords AB and CD be drawn. Then AC and BD cut l at equal distances from P, as do also AD and BC.*

This simple appearing theorem is surprisingly difficult to prove. It is a special case of a rather more general theorem, which we may state and prove at once.

Theorem. *Given a complete quadrangle inscribed in a circle; if any line cuts two opposite sides at equal distances from the center of the circle, it cuts each pair at equal distances from the center.*

Proof:* Let A, B, C, D be on a circle with center O, and

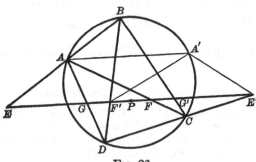

FIG. 26

let AB, CD, AC, BD, AD, BC meet a line XY at E, E', F, F', G, G' respectively. Let P be the foot of the perpendicular from O to XY. If now $\overline{OE} = \overline{OE'}$, which is the same as $\overline{PE} = \overline{PE'}$, we are to prove $\overline{PF} = \overline{PF'}$ and $\overline{PG} = \overline{PG'}$. Drawing AA', a chord of the circle parallel to XY, we see that $AA'E'E$ is an isosceles trapezoid; by equal angles we find that F', E',

---

* This proof is due to Mackay, *Proceedings of Edinburgh Math. Society*, III, 1884, p. 38; the theorem is the point of departure of a remarkable article by A. L. Candy, *Annals of Math.*, 1896, p. 175. The reader who is acquainted with projective geometry will recognize a familiar theorem on involutions.

A', D lie on a circle, so that angles EAF and $F'A'E$ are equal, and triangles EAF and $E'A'F'$ are congruent.

106. Theorem. *Let AB and $A'B'$ be parallel but not equal, and let AA' meet BB' at P, and AB' meet $A'B$ at Q.*

Let

$$\frac{\overline{PA'}}{\overline{PA}} = \frac{\overline{PB'}}{\overline{PB}} = \frac{m}{n}$$

Then

$$\frac{\overline{A'Q}}{\overline{A'B}} = \frac{\overline{B'Q}}{\overline{B'A}} = \frac{m}{m+n}$$

For triangles ABQ and $B'A'Q$ are similar, hence

$$\frac{\overline{A'Q}}{\overline{QB}} = \frac{\overline{B'Q}}{\overline{QA}} = \frac{m}{n}$$

whence the desired result by composition. This theorem is a lemma for the following rather more interesting one.

Theorem. *Given in the plane $k + 1$ points, $A_1, A_2, \ldots A_k$, P, and a fraction m/n. If we construct a broken line $PP_1P_2P_3 \ldots P_k$, laying off $\overline{PP_1}$ along $\overline{PA_1}$ and equal to m/n of it; then each successive segment toward one of the A's, with multipliers in harmonic progression, viz:*

$$\overline{PP_1} = \frac{m}{n} \overline{PA_1}, \overline{P_1P_2} = \frac{m}{m+n} \overline{P_1A_2}, \overline{P_2P_3} = \frac{m}{2m+n} \overline{P_2A_3}, \text{etc.,}$$

then the same point P is the end of the broken line, in whatever order the points $A_1, A_2, \ldots A_k$ are taken.

If there are only two points the preceding theorem establishes the result. If there are more than two, any change of order can be effected by a succession of interchanges of pairs of A's, no one of which affects the result.

107. We close the chapter with a few theorems concerning areas.

Theorem. *Two triangles whose vertices lie on the sides of a given triangle at equal distances from their mid-points are equal in area.*

That is, if P_1 and Q_1 lie on the side A_2A_3 of triangle $A_1A_2A_3$,

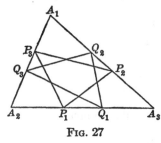

FIG. 27

so that $\overline{A_2P_1} = \overline{Q_1A_3}$, and if P_2, Q_2, P_3, Q_3 are similarly placed, then area $P_1P_2P_3$ equals area $Q_1Q_2Q_3$.

For let

$$\overline{A_2P_1} = \overline{Q_1A_3} = m_1,$$
$$\overline{A_3P_2} = \overline{Q_2A_1} = m_2,$$
$$\overline{A_1P_3} = \overline{Q_3A_2} = m_3$$

Since the areas of two triangles having a common angle are proportional to the products of the including sides,

$$\text{area } A_1P_2P_3 = \frac{(a_2 - m_2)\, m_3}{a_2a_3}\, \Delta, \text{ etc.}$$

But $P_1P_2P_3 = \Delta - A_1P_2P_3 - A_2P_3P_1 - A_3P_1P_2$

$$= \Delta\left[1 - \frac{(a_2 - m_2)\, m_3}{a_2a_3} - \frac{(a_3 - m_3)\, m_1}{a_3a_1} - \frac{(a_1 - m_1)m_2}{a_1a_2}\right]$$

$$= \Delta\left[1 - \left(\frac{m_1}{a_1} + \frac{m_2}{a_2} + \frac{m_3}{a_3}\right) + \frac{m_2m_3}{a_2a_3} + \frac{m_3m_1}{a_3a_1} + \frac{m_1m_2}{a_1a_2}\right]$$

Working out the area of $Q_1Q_2Q_3$ by exactly the same method, we obtain the same formula for it.

A case of special interest, which will arise again later (**276**, **476** ff.), occurs when the three sides are divided in the same ratio.

108. Theorem. *If triangle $P_1P_2P_3$ is inscribed in $A_1A_2A_3$, and if P_1Q_2 is drawn parallel to A_2A_1, meeting A_1A_3 at Q_2, etc., then triangles $P_1P_2P_3$ and $Q_1Q_2Q_3$ have equal areas. In particular, if P_1, P_2, P_3 are collinear, and if P_1Q_2, etc., are drawn as before, then Q_1, Q_2, Q_3 are collinear.*

109. Theorem.* *If the sides of a convex quadrilateral inscribed in a circle, taken in order, are a, b, c, d, and if s denotes half the perimeter, the area of the quadrilateral is given by*

$$F = \sqrt{(s-a)\ (s-b)\ (s-c)\ (s-d)}$$

It is to be noted that this is an extension of a familiar result; if we take $d = O$, we have the common formula for the area of a triangle.

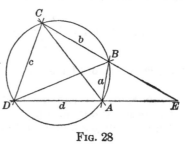

Proof: let the figure be $ABCD$, where $\overline{AB} = a$, etc. If it is a rectangle, the proof is immediate. If not, let BC and AD meet at E, outside the circle. Denoting $\overline{CE}$ by x, $\overline{DE}$ by y, we have

FIG. 28

$$\text{area } CDE = \tfrac{1}{4}\sqrt{(x+y+c)\ (x+y-c)\ (x-y+c)\ (-x+y+c)}$$

But triangles ABE and CDE are similar, and

$$\frac{\text{area } ABE}{\text{area } CDE} = \frac{a^2}{c^2}$$

Treating this proportion by division, in other words subtracting each side of the equation from unity,

$$\frac{\text{area } ABCD}{\text{area } CDE} = \frac{c^2 - a^2}{c^2}$$

Further, we have the proportions

$$\frac{x}{c} = \frac{y-d}{a}, \qquad \frac{y}{c} = \frac{x-b}{a}$$

---

* This and the following theorems, dealing primarily with the problem of the area of a simple quadrangle inscribed in a circle, are taken, with one or two exceptions, from Fuhrmann, *l.c.*, pages 75–78. They seem to be less well known than they merit.

Adding these and solving for $x + y$, we may obtain

$$x + y + c = \frac{c}{c - a} (a + b + d - c)$$

Similar expressions for $x + y - c$, etc., are found at once. Substituting and reducing, we find

$$\text{area } CDE = \frac{c^2}{c^2 - a^2} \sqrt{(s - a)(s - b)(s - c)(s - d)}$$

whence the desired result.

Extension. It can be proved that for any convex quadrilateral having sides a, b, c, d, and the sum of one pair of opposite angles equal to $2u$, the area Δ is given by

$$\Delta^2 = (s - a)(s - b)(s - c)(s - d) - abcd \cos^2 u$$

We shall not take up the proof, as it involves long and rather unpleasant trigonometric reductions. From this formula it is immediately clear that of all the quadrilaterals that can be formed with four given lines as sides, the largest is that one inscribed in a circle; a fact that is of course demonstrable by more elementary methods. The elementary texts, however, overlook the necessity of proving the existence of such a cyclic quadrangle. We now proceed to furnish a proof * that there exists a quadrangle whose sides are equal to those of a given quadrangle, and whose vertices lie on a circle.

Problem. *To construct a cyclic quadrangle, given the lengths of the sides in order.*

Let a, b, c, d be four given lengths; and suppose that the completed figure is $ABCD$, with $\overline{AB} = a$, $\overline{BC} = b$, $\overline{CD} = c$, $\overline{DA} = d$. Draw CM so that $\angle DCM = \angle CAB$, meeting

---

* McClelland, *l.c.*

AD at M. Then triangles CDM and ABC are similar; and DM, being a fourth proportional to the given lines a, b, c, can be constructed from the data of the problem.

The construction, then, is effected by laying off segments AD equal to d, and DM equal to $\dfrac{bc}{a}$. Now we can determine two loci for the point c; first, CD shall be equal to c, hence we construct a circle of radius c about the known point D. Second, since

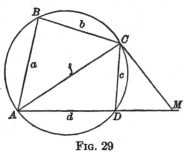

$$\frac{\overline{AC}}{\overline{CM}} = \frac{a}{c}$$

FIG. 29

a locus for c is a circle which can be constructed (**58**). By a long algebraic computation which presents no difficulty, we find that these two circles intersect, provided each of the four lines is less than the sum of the other three; in other words, provided that any quadrilateral whatever is possible with the given sides. It follows that the point C can be located, and we have the solution, which is essentially unique.

110. Suppose now that we have the same four lines, in a different order. We observe that the quadrilateral may be inscribed in the same circle that we have just found. And since the solution is unique, we may make the following statements:

Theorem. *Given four lengths, a, b, c, d, each of which is less than the sum of the other three. In any one of the three possible cyclic orders, they may be taken as the sides of a cyclic quadrangle in one and only one way. The three quadrangles thus determined are not in general similar; but their circumscribing circles are equal, and the three quadrangles have the same area, namely*

$$F = \sqrt{(s-a)\,(s-b)\,(s-c)\,(s-d)},$$

where $s = \frac{1}{2}\,(a+b+c+d).$

Any two of the three quadrangles have a diagonal of one equal to a diagonal of the other.

111. The last remark is suggestive. The six diagonals of the three quadrilaterals are equal in pairs. Let us designate by e the line which separates a and d from b and c; by f, that which separates a and b from c and d; and by g that which separates a and c from b and d. With reference to these three lines, we have a remarkable theorem.

Theorem. *Let a quadrilateral be inscribed in a circle of radius R, and let its three diagonals, in the sense of the foregoing paragraph, be e, f, g. Then the area of the quadrilateral is*

$$F = \frac{efg}{4R}$$

The proof is based on the known formula (**15** *d*) for the area of the triangle; the areas bounded by a, b, f and by c, d, f (fig. 29) are respectively

$$F_1 = \frac{abf}{4R} \quad \text{and} \quad F_2 = \frac{cdf}{4R}$$

Adding, and applying the theorem of Ptolemy,

$$F = \frac{f}{4R}\,(ab+cd) = \frac{feg}{4R}$$

Corollary. We incidentally obtain formulas for the lengths of the diagonals in terms of the sides:

$$f^2 = \frac{(ac+bd)\,(ad+bc)}{(ab+cd)}, \text{ etc.}$$

For $F = \dfrac{f}{4R}\,(ab+cd) = \dfrac{g}{4R}\,(ac+bd) = \dfrac{e}{4R}\,(ad+bc)$

whence

$$\frac{f}{g} = \frac{ac + bd}{ab + cd} \qquad \frac{f}{e} = \frac{ad + bc}{ab + cd} \qquad eg = ab + cd$$

Multiplying together these three, we obtain at once the result. From these expressions for e, f, and g, we may in turn, with the aid of **109**, work out a formula for R in terms of the sides.

Exercise. Furnish complete proofs of all propositions in the chapter whose proofs are omitted in whole or in part, viz: **85, 86, 87, 93, 94, 96** (including corollaries), **97, 98, 99, 101** (including corollaries), **102, 103, 104, 108**.

CHAPTER V

GEOMETRY OF CIRCLES

112. In Chapter III we have studied the essentials of the geometry of the circle, with especial emphasis on the properties of coaxal circles and of the inversion transformation. In the present chapter we carry somewhat further the same studies. We add to our stock of working methods some additional tools nearly as important as those introduced in the earlier chapter. These methods will not be extensively used in the sequel, and the reader may without serious embarrassment omit Chapters V and VI entirely, passing at once to the geometry of the triangle in Chapter VII. It is strongly recommended, however, that the following portions be read; **113–117, 126–133.** Chapter VI is a further study in circles and is in no sense prerequisite for the later work.

The first portion of this work is based on an important theorem of Casey relating to coaxal circles; from this central theorem we are led to consider several interesting extensions. In the theory of inversion, we next develop the properties of the "circles of antisimilitude" with regard to which two given circles are mutually inverse. We pass to a brief discussion of poles and polars, a topic intimately related to inversion, and finally to a special form of inversion in space, known as stereographic projection.

113. The following theorem, due to Casey, is an open sesame to a number of theorems and developments.

Theorem. *The difference of the powers of a point with regard to two non-concentric circles is twice the product of the distance between their centers by the distance from the point to the radical axis of the circles* (**45, 98**).

The proof closely resembles that of the radical-axis theorem (**45**).   Let the given circles be $C_1(r_1)$, $C_2(r_2)$, P the given point, PQ and PP' the perpendiculars from P to the radical axis LQ and the line of centers C_1C_2LP' respectively.   We have then

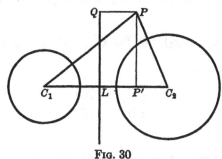

FIG. 30

Difference of powers $= \overline{PC_1}^2 - r_1^2 - \overline{PC_2}^2 + r_2^2$

$$= \overline{P'C_1}^2 - (\overline{C_1C_2} + \overline{P'C_1})^2 - r_1^2 + r_2^2$$

$$= 2\,\overline{C_1C_2}\cdot\overline{P'C_1} - \overline{C_1C_2}^2 - r_1^2 + r_2^2$$

$$= 2\,\overline{C_1C_2}\cdot\overline{P'C_1} + 2\,\overline{C_1C_2}\cdot\overline{C_1L} \qquad (45)$$

$$= 2\,\overline{C_1C_2}\cdot\overline{PQ}$$

Corollary.   *The locus of a point, the difference of whose powers with regard to two circles is constant, is a straight line parallel to their radical axis.*

Corollary.   *If a point moves on a circle, its power with regard to a second circle is proportional to its distance from the radical axis of the two circles.*   (The factor of proportionality is equal to twice the distance between the centers of the circles.)

This is the same as the foregoing theorem, when the power with regard to one of the circles is zero.

114. Theorem.   *If a point moves on one circle of a coaxal system, the ratio of its powers with regard to two other circles* *of the system is constant, and is equal to the ratio of the distances between the corresponding centers.*

Let the three coaxal circles be c, c_1, c_2; and let P be any point of c. Denoting the perpendicular from P to the radical axis by PQ, and the power of P with regard to any circle c by $P(c)$, we have

$$P(c) = 0, \quad P(c_1) = 2\,\overline{PQ} \cdot \overline{CC_1}, \quad P(c_2) = 2\,\overline{PQ} \cdot \overline{CC_2}$$

whence immediately

$$\frac{P(c_1)}{P(c_2)} = \frac{\overline{CC_1}}{\overline{CC_2}}$$

as was to be proved.

115. Theorem. *Conversely, the locus of a point whose powers with regard to two fixed circles are in a constant ratio is a circle coaxal with them.*

For let P be any position of the point satisfying the condition; and let X be the center of the circle which is coaxal with the given circles and passes through P. Then

$$P(c_1) = 2\,\overline{C_1X} \cdot \overline{PQ}, \quad P(c_2) = 2\,\overline{C_2X} \cdot \overline{PQ}$$

so that

$$\frac{P(c_1)}{P(c_2)} = \frac{\overline{C_1X}}{\overline{C_2X}}$$

The left-hand member is constant by hypothesis; hence X is a fixed point of C_1C_2, and P is always on the same circle of the coaxal system.

As a special case, we have already noted that if a point moves on a circle of a coaxal system its distances from the limiting points of the system are in a constant ratio.

Theorem. *The circle of similitude of two circles is coaxal with them.* (Cf. **37, 59**.)

116. The theorem of Casey, as stated, is not applicable to concentric circles. We have the following substitute, which enables us to establish the theorems of **114** and **115** for concentric circles.

Theorem. *The difference between the powers of a point with regard to two concentric circles equals the difference of the squares of the radii, and is everywhere constant.*

Theorem. *The locus of a point whose powers with regard to two concentric circles are in a constant ratio, is a circle concentric, i.e., coaxal with them.*

We proceed to some applications of this group of theorems.

117. Theorem. *If AP, BQ, CR are the tangents to a circle K from three points A, B, C, the circle through A, B, C is tangent to K if and only if*

$$\overline{AB}\cdot\overline{CR} \pm \overline{AC}\cdot\overline{BQ} \pm \overline{BC}\cdot\overline{AP} = 0$$

We recognize this as of the same form as Ptolemy's theorem; when K is a null-circle this theorem reduces to that of Ptolemy. In the next chapter we shall have a further generalization concerning the common tangents to four circles (**172**).

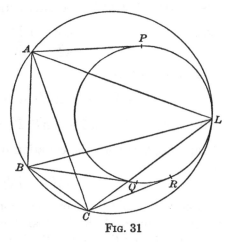

Proof: designating by J the circle through the points A, B, C, we first assume that this circle is tangent to K at a point L. Hence we may regard L as a null-circle and the three circles J, K, L are members of a coaxal system of the third type. Therefore, by **113**,

FIG. 31

second corollary, as a point moves on J its powers with regard to K and L are in a constant ratio:

$$\overline{AP} = c \cdot \overline{AL}, \quad \overline{BQ} = c \cdot \overline{BL}, \quad \overline{CR} = c \cdot \overline{CL}$$

But A, B, C, L are concyclic. Supposing that B is opposite L, we have by Ptolemy's theorem

$$\overline{BL} \cdot \overline{AC} = \overline{AL} \cdot \overline{BC} + \overline{CL} \cdot \overline{AB}$$

Multiplying through by c and substituting, we have

$$\overline{BQ} \cdot \overline{AC} = \overline{AP} \cdot \overline{BC} + \overline{CR} \cdot \overline{AB}$$

as was to be shown. Conversely, we shall now assume that this equation is true, and prove that the circles are tangent.

The locus of a point X, such that

$$\frac{\overline{AX}}{\overline{CX}} = \frac{\overline{AP}}{\overline{CR}}$$

is a circle (58); this circle cuts the circle ABC once on each side of AC. Let M be the intersection opposite to B; then

$$\frac{\overline{AM}}{\overline{AP}} = \frac{\overline{CM}}{\overline{CR}} = t$$

By Ptolemy's theorem,

$$\overline{BM} \cdot \overline{AC} = \overline{AM} \cdot \overline{BC} + \overline{CM} \cdot \overline{AB}$$

Substituting, $\overline{BM} \cdot \overline{AC} = t \cdot \overline{AP} \cdot \overline{BC} + t \cdot \overline{CR} \cdot \overline{AB}$

Comparing this with the equation of the hypothesis, we see that

$$\overline{BM} = t \cdot \overline{BQ}$$

so that B is also a point on the same circle ACM. That is, the circle through A, B, C is coaxal with K and the null-circle M. But M is on the circle ABC, hence the coaxal system is of the third type, and all its members are tangent at M.

118. Theorem. *The tangents to two circles at any four collinear points intersect at four points on a circle coaxal with the given circles.*

For let a line cut a circle at P and Q, and a second circle at R and S. Let the tangents to the circles at P and R, for instance, meet at A; then the ratio of the tangents to the circles from A is

$$\frac{\overline{AP}}{\overline{AR}} = \frac{\sin \angle APQ}{\sin \angle ARS}$$

Since the tangents to either circle make equal angles with the secant, it follows that for the four points this ratio has a constant value, so that by **115** the points are on a circle coaxal with the given circles.

A special case occurs when the line goes through a center of similitude of the circles; two of the points of intersection are at infinity, and the other two are on the radical axis.

Theorem. *If tangents are drawn to two circles from a moving point on a circle coaxal with them, the line through the points of tangency cuts the circles in chords having a constant ratio. In particular, if a line cuts two circles in equal chords, the tangents at the points of intersection meet on the circle of similitude, and conversely.*

119. The theorems of Poncelet concerning polygons inscribed and circumscribed to circles furnish an interesting application of the foregoing power theorems.

Lemma. *If the vertices of a complete quadrangle are on a circle, a transversal that cuts two opposite sides at equal angles cuts each pair at equal angles.*

Let A, B, C, D lie on a circle, and let a line XY cut the circle at X and Y and meet AB and CD at equal angles at P and Q respectively. Then

$$\angle AB, XY = \angle XY, CD$$

But $$\angle BD, AB = \angle CD, AC$$

Adding, $$\angle BD, XY = \angle XY, AC$$

as was to be proved.

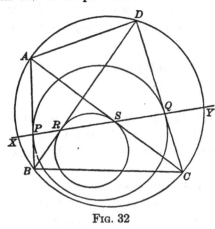

Fig. 32

Lemma. *If a line makes equal angles with the opposite sides of a cyclic quadrangle, a circle can be drawn tangent to each pair where this line meets them; and these three circles are coaxal with the given circle.*

Since the members of each pair make equal angles with the transversal, a circle can be drawn tangent to them at the intersections. Consider now two of these circles; we have the tangents to these circles at collinear points, as in **118**; hence the intersections of these tangents, namely A, B, C, D, lie on a circle coaxal with the two circles before us. It follows that the four circles are coaxal.

120. Theorem. *If a quadrangle inscribed in a fixed circle moves so that two opposite sides remain tangent to a fixed circle, any pair of opposite sides are tangent in each position to some circle coaxal with the two fixed circles.*

121. Theorem. *If a triangle moves continuously with its vertices on one circle of a coaxal system, while two of its sides continuously touch other fixed circles of the system, then the third side touches a fixed circle of the system.*

* As this theorem is ordinarily stated, without the stipulation of continuity, it is not true. If we postulate that the triangle be inscribed in a circle while two of the sides are tangent to other circles, then the third side is tangent to one or the other of two different circles.

Let $A_1A_2A_3$ and $B_1B_2B_3$ be two positions of a triangle inscribed in a circle c of a coaxal system, while A_1A_2 and B_1B_2 are tangent to a circle c_3, A_1A_3 and B_1B_3 to a circle c_2. We wish to prove that A_2A_3 and B_2B_3 are tangent to another circle c_1 of the same coaxal system. Consider the quadrangle $A_1B_1A_2B_2$; since two of its opposite connectors are tangent to a circle, so also are another pair, namely A_1B_1 and A_2B_2. Similarly, A_1B_1 and A_3B_3 are tangent to a circle of the system. Now in a coaxal system there are generally two circles tangent to a given line, as A_1B_1; and we have to determine whether the circle c' which touches A_1B_1 and A_2B_2 is the same as c'', which touches A_1B_1 and A_3B_3, or distinct. By the principle of continuity we can show that they are identical. For let $B_1B_2B_3$ move continuously into $A_1A_2A_3$; then c' and c'' both move continuously into the given circle c, while A_1B_1 and A_2B_2 move into the positions of the tangents to c at A_1 and at A_2 respectively. The tangent to c at A_1 is of course tangent also to a second circle $\bar{c}$ of the coaxal system, which is on the opposite side of the radical axis. As B_1 moves into the position A_1, there are always two circles of the coaxal system tangent to A_1B_1; one of them moves into the limiting position c, the other into $\bar{c}$. But both c' and c'' move into c, not into $\bar{c}$, hence these two are at all times identical.

Hence we have a circle of the system touching A_2B_2 and A_3B_3. Therefore, by the same argument used before, there is a circle c_1 tangent to A_2A_3 and B_2B_3. But A_2A_3 is a fixed line, and as B_2B_3 moves continuously it can be tangent to only one fixed circle. This completes the proof.

122. Theorem. *If a polygon moves with its vertices on a fixed circle, and if each of its sides with one exception is known to touch a fixed circle of a coaxal system including the first circle, then the remaining side touches a fixed circle of the system, and each of the diagonal connectors touches a circle of the system.*

This is an immediate consequence of the foregoing theorem. Whenever two lines from a vertex are known to touch circles of the system, the line joining their extremities also touches a circle.

In particular, it may happen that all the sides of the polygon touch one and the same circle. We then have the following theorem.

Theorem. *If two circles are so related that a polygon can be inscribed to one and circumscribed to the other, then infinitely many polygons can be so drawn, and each diagonal of the variable polygon is tangent to a fixed circle.*

123. The problems relating to a triangle inscribed to one circle and circumscribed to another will be considered in due order in **297.** At this time we may consider briefly the quadrilateral so constructed.

Theorem. *If a moving chord of a circle subtends a right angle at a fixed point M, the mid-point of the chord and the foot of the perpendicular on it from M trace one and the same circle; the point of intersection of the tangents at the ends of the chord also traces a circle; and the three circles are coaxal, with M as one limiting point of the system.*

For let AB be the moving chord, O its mid-point, H the foot of the perpendicular; and let the tangents to the given circle at A and B meet at P. Then

$$\overline{OM} = \overline{OA} = \overline{BO}$$
$$\overline{OM}^2 = - \overline{OA} \cdot \overline{OB}$$

and the powers of O with regard to the given circle and the null-circle M are in a constant ratio -1, whence O moves on a circle coaxal with these two. Also by similar triangles,

$$\overline{HM}^2 = - \overline{HA} \cdot \overline{HB},$$

and H is always on the same circle. Finally, O and P are inverse points with regard to the circle, hence the locus of P is another circle of the same system.

124. Theorem. *If a quadrilateral is circumscribed about a circle, its vertices lie on another circle if and only if the lines joining the points of contact of opposite sides are mutually perpendicular.*

For let the sides of a quadrilateral $A_1A_2A_3A_4$ be tangent to a circle c at B_1, B_2, B_3, B_4; if B_1B_3 and B_2B_4 are mutually perpendicular at a point M, then each of the chords B_1B_2, B_2B_3, B_3B_4, B_4B_1 subtends a right angle at M, and by the previous theorem A_1, A_2, A_3, A_4 lie on a circle. Conversely, if these four points are assumed to lie on a circle, we can easily prove by equal arcs that B_1B_3 and B_2B_4 meet at right angles.

Corollary. *If two circles are so situated as to admit a quadrilateral inscribed in one and circumscribed to the other, they admit infinitely many; any point of the first circle may be taken as a vertex.*

For under these circumstances, wherever the point A_1 is located, if the tangents A_1B_1 and A_1B_4 are drawn, then B_1B_4 will subtend a right angle at M.

125. Theorem. *If r and ρ are the radii of two circles admitting in- and circumscribed quadrilaterals, and d the distance between their centers, then*

$$\frac{1}{(r-d)^2} + \frac{1}{(r+d)^2} = \frac{1}{\rho^2}$$

For let the line of centers cut the first circle in A_1 and A_3; let the sides of the quadrilateral $A_1A_2A_3A_4$ be tangent to the second circle at B_1, B_2, B_3, B_4, so that B_1B_4 and B_2B_3 are perpendicular to the line of centers OC, say at D and E. Then, since $A_1A_2A_3$ is a right angle, B_1O is perpendicular to B_2O, and triangles ODB_1 and B_2EO are congruent.

But $\qquad\qquad \overline{OD}^2 + \overline{DB_1}^2 = \rho^2$

hence $\qquad\qquad \overline{OD}^2 + \overline{OE}^2 = \rho^2$

Since $\quad OD = \dfrac{\rho^2}{r+d}$ and $OE = \dfrac{\rho^2}{r-d}$

we get the desired equation after reducing.

CIRCLES OF ANTISIMILITUDE

126. Definition. A circle of antisimilitude of two circles is a circle with regard to which they are mutually inverse.

We have already proved that the center of inversion is a center of similitude for any two mutually inverse circles, and that corresponding points on these circles are antihomologous. Hence two given circles have at most two circles of antisimilitude, which in all cases are coaxal with them. In order that a center of similitude of two circles may be the center of a circle of antisimilitude, it is necessary and sufficient that the constant product of the distances to antihomologous points be a positive number; then this number is the square of the radius of inversion. Viewing all the cases, we are enabled to state the following results.

127. Theorem. *Two intersecting circles have two circles of antisimilitude, passing through the points of intersection and mutually orthogonal, with their centers at the centers of similitude of the given circles. Two non-intersecting or tangent circles have a single circle of antisimilitude, which is coaxal with them and whose center is the external or the internal center of similitude, according as the circles are mutually external or one is inside the other.*

128. If the given circles are transformed by inversion into concentric circles or straight lines, we may deduce the results just given by referring to the simpler figure.

Theorem. *Two concentric circles have a single circle of antisimilitude, which is concentric with them and whose radius is the mean proportional of their radii. Two intersecting straight lines have two circles of antisimilitude, their angle bisectors, which are mutually orthogonal.*

129. Theorem. *Any two circles can be transformed by inversion into equal circles.*

It is sufficient to place the center of inversion on either circle of antisimilitude; the latter then inverts into a straight line, and circles inverse with regard to it are equal and symmetrical.

Theorem.* *Given three circles, there exist not more than eight points, any one of which being taken as center of inversion, the transformed circles are equal. There may, however, be no such points.*

The result will be achieved if the center of inversion is an intersection of circles of antisimilitude of any two pairs of the given circles. Through such a point obviously passes a circle of antisimilitude of the third pair. In the most favorable case, when all the circles intersect and there are three pairs of circles of antisimilitude, those of any two pairs will intersect in eight points; on the other hand, it may well happen that no two circles of antisimilitude intersect. This occurs, for instance, if one circle be very large, while the others are relatively small and at a great distance from the first and from each other.

This negative result is highly regrettable, for it would be a real advantage to be able in all cases to invert three given circles into equal circles. In the special case that the given circles are concurrent, this transformation is always possible.

Corollary. *There exist at most eight centers of inversion,*

* Incorrect versions of this theorem are frequent; cf. Lachlan, p. 223, and Casey, p. 90. The statement enunciated by McClelland, p. 246, is correct.

with regard to which three given circles invert into circles with any chosen radii. There may, however, be no such point.

130. Theorem. *There exists in general an inversion by which the inverses of three given points are vertices of a triangle similar to a second given triangle.*

Suppose triangles ABC and PQR are given, and it is desired to effect an inversion of A, B, C into three points A', B', C' such that triangle $A'B'C'$ is similar to PQR; then by **75**, the center of inversion O is determined by the equations

$$\angle\, AOB = \angle\, ACB + \angle\, PRQ$$
$$\angle\, BOC = \angle\, BAC + \angle\, QPR$$

as the intersection of two circles, the former through A and B, the latter through B and C. (See, however, **75**, remarks.)

Corollary. *Any two triangles may be so placed that their vertices are mutually inverse with regard to a circle.*

131. Theorem.* *Any four points not on a circle may be inverted into the vertices and orthocenter of a triangle.*

To prove this theorem, we first consider the circles of antisimilitude of three circles ABC, ABD, ACD through a point A. If we invert these three circles into straight lines, $B'C'$, $B'D'$, $C'D'$, the circles of antisimilitude are transformed into the bisectors of the angles of triangle $B'C'D'$. These angle bisectors are concurrent at four points, therefore in the original figure the six circles of antisimilitude are concurrent at four points. Now let one of these points be taken as a center of inversion. The three circles of antisimilitude through it are transformed into straight lines, hence the circles ABC, ABD, ACD are transformed into equal circles. Whence as indicated in **104** the intersections A'', B'', C'', D'' of these equal

---

* Concerning this and the following theorems, cf. Johnson, "On the Circles of Antisimilitude of the Circles determined by Four Given Points," *American Mathematical Monthly*, XXX, 1923, p. 250.

circles have the property that any one is the orthocenter of the triangle of the other three.

132. Theorem. *Any four points can be inverted into the vertices of a parallelogram.*

If the four points are not on a circle, we consider, as before, the circle through each three of them, and the circles of antisimilitude of these four circles. Besides the four points already determined, where these are concurrent, the circles of antisimilitude of ABC and ADC will meet those of ABD and CBD at four other points. (Consideration of the figure as simplified by inversion will demonstrate the real existence of these intersections.) If such a point is taken as center of inversion, the resulting circles will be equal in pairs and their intersections will be vertices of a parallelogram.

Theorem. *Any four points on a circle can be inverted into the vertices of a rectangle.*

Let A and C separate B and D on a circle. The circles orthogonal to the given circle, the one at A and C and the other at B and D, will intersect at two points X and Y. Let X be a center of inversion, then the circles ACX and BCX invert into lines intersecting at Y', and the given circle into a circle orthogonal to them, with its center therefore at Y'. Hence $A'C'$ and $B'D'$ are diameters, and the figure is a rectangle.

133. Some of the foregoing theorems, and others of the sort, are intimately related to the following general theorem.

Theorem. *If two polygons are inscribed in the same circle, and the connectors of corresponding vertices are concurrent at a point C, the inverse of either polygon with regard to C as center of inversion is homothetic to the second polygon.*

For as we saw in the proof of **71**, the mutually inverse points are antihomologous on the given circle and its inverse. Hence the points here in question are homologous, with C as homothetic center.

Numerous applications of this theorem are obvious. The form of the triangle inverse to a given triangle is at once determined when the center of inversion is specified. In the proof of the second theorem of **132,** if either X or Y be connected with the four given points, the connectors meet the circle again at the vertices of a rectangle. As another application, we may consider the harmonic quadrangle.

Definition. Any quadrangle whose vertices are inverse to those of a square is a *harmonic quadrangle.*

Theorem. *A cyclic quadrangle is harmonic if and only if the products of its opposite sides are equal.*

For this is a property of a square, and of no other rectangle; and by **68** *c* the property is unchanged by an inversion.

Theorem. *Lines from the vertices of a square through any point cut the circumscribed circle in the vertices of a harmonic quadrangle.*

POLES AND POLARS

134. The theory of poles and polars belongs properly to the domain of projective geometry, and can hardly be adequately treated by elementary methods. But in view of its rather close association with inversion we give it some brief consideration.

Definition. If two points are inverse with regard to a circle, the straight line through the second which is perpendicular to the line of the points is called the *polar* of the first with regard to the circle. The point is called the *pole* of the line.

135. The following properties are immediate consequences of the definition.

Theorem. *Every point except the center of inversion has a definite polar, and any line not passing through the center has a pole. The polar of a point on the circle of inversion is the tangent to that point, and the pole of a tangent is the point of contact; in no other case does a polar pass through its pole. If a point is outside the circle, its polar is the line joining the points of contact of the tangents from it to the circle. The angle between two lines equals the angle subtended at the center of inversion by their poles.*

For completeness, the pole of the center of inversion is defined to be the line at infinity; and the pole of any diameter of the circle of inversion is the point at infinity in the direction perpendicular to it.

136. Theorem. *If a point lies on the polar of a second, the second is on the polar of the first.*

For let Q lie on the polar of P; and denote the inverses of these points by Q' and P', the center of inversion by O. Then the polar of P is perpendicular to OPP' at P', and $OP'Q$ is a right triangle. But triangles $OP'Q$ and $OQ'P$ are similar, hence P is on the perpendicular to OQ' at Q', which is the polar of Q.

Hence, *if several points are collinear, their polars are concurrent; and if several lines are concurrent, their poles are collinear. The pole of the line joining two points is the point of intersection of the polars of the points.*

137. Theorem. *If secants are drawn to a circle through a fixed point, the point of intersection of the tangents where any secant meets the circle lies on the polar of the given point.*

That is, if through a fixed point A we draw a line meeting the circle at P and Q, the tangents at P and Q meet at a point T on the polar of A. For the polar of T is the line PQ; and since PQ passes through A, then T lies on the polar of A.

Exercise: State and prove the converse theorem.

138. Definition. Two points, each of which lies on the polar of the other, are called *conjugate points* with regard to the circle; and two lines, each of which passes through the pole of the other, are called *conjugate lines.*

Theorem. *If the line joining two conjugate points meets the circle, either pair of points divides the other externally and internally in the same ratio; that is, any two conjugate points, with the points where their line meets the circle, form a harmonic set (87). Conversely, any two points which divide harmonically a secant of a circle are conjugate with regard to it.*

Let O be the center of the circle, P and Q any conjugate points. Let PQ cut the circle at X and Y, and let the circle

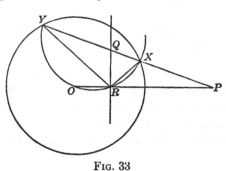

Fig. 33

XYO cut OP at R. Then triangles OXP and ORX are similar, and therefore R is the inverse of P, RQ is its polar, and PRQ is a right angle. In triangle XYR, RP bisects the exterior angle R; therefore RQ, which is perpendicular to it, bisects the interior angle. We know, however, that the bisectors of an angle of a triangle divide the opposite side in the ratio of the adjacent sides, and therefore P and Q divide XY internally and externally in the same ratio, as was to be proved.

Corollary. *The locus of the harmonic conjugates of a given point with regard to the intersections of the circle of reference with variable secants through the point, is the polar of the point.*

139. Theorem. *If from a fixed point two secants are drawn*

---

* This is a projective theorem, and most proofs depend on projective principles. The above simple proof is adapted from Lachlan (p. 152).

to a circle, and their extremities are connected in pairs, the opposite connectors intersect on the polar of the point.

From A let lines APQ and ARS be drawn to a circle; let PS meet QR at Y, and PR meet QS at Z. We wish to prove that Y and Z are on the polar of A.

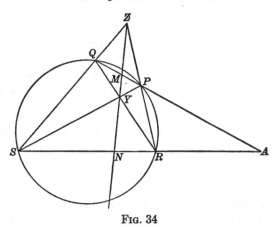

Fɪɢ. 34

A simple proof of this theorem can be based on Chapter VIII (cf. **225**); the following proof, though apparently formidable, is a straightforward application of **84**.

Let YZ meet APQ at M, and ARS at N. It is sufficient to show that

$$\frac{\overline{MP}}{\overline{MQ}} = -\frac{\overline{AP}}{\overline{AQ}} \quad \text{and} \quad \frac{\overline{NR}}{\overline{NS}} = -\frac{\overline{AR}}{\overline{AS}}$$

We have by **84**

$$\frac{\overline{MP}}{\overline{MQ}} = \frac{\overline{ZP}\sin\angle MZP}{\overline{ZQ}\sin\angle MZQ} \quad \text{and} \quad \frac{\overline{YQ}}{\overline{YR}} = \frac{\overline{ZQ}\sin\angle YZQ}{\overline{ZR}\sin\angle YZR}$$

$$\frac{\overline{ZP}}{\overline{ZR}} = \frac{\overline{SP}\sin\angle QSP}{\overline{SR}\sin\angle QSR} \quad \text{and} \quad \frac{\overline{YR}}{\overline{YQ}} = \frac{\overline{SR}\sin\angle PSR}{\overline{SQ}\sin\angle PSQ}$$

Combining and canceling,

$$\frac{\overline{MP}}{\overline{MQ}} = -\frac{\overline{SP}\sin\angle PSR}{\overline{SQ}\sin\angle QSR}$$

But similarly, by direct application of **84**

$$\frac{\overline{AP}}{\overline{AQ}} = \frac{\overline{SP}\sin\angle PSR}{\overline{SQ}\sin\angle QSR}$$

whence the desired result for M at once. The same method may be applied at once to N. Hence MN is the polar of A.

Corollary. *The polar of a point with regard to a circle can be constructed with ruler only, by means of a complete quadrangle inscribed in the circle.*

140. Problem. *To draw the tangents to a circle from an outside point.*

We draw the polar by the method indicated above, and the tangents are the lines from the given point to the intersections of circle and polar. This construction with the ruler only is frequently used in practice.

141. Theorem. *The circle having as extremities of a diameter two points conjugate with regard to a given circle is orthogonal to the latter. Conversely, if two circles are orthogonal, the extremities of any diameter of one are conjugate with regard to the other.*

Corollaries. *a. If P is a fixed point of a given circle, the polars of P with regard to all circles orthogonal to the given circle pass through the fixed point diametrically opposite to P.*
b. The distance between two conjugate points is twice the tangent to the circle from the point midway between them.
c. The polars of a fixed point with regard to the circles of a coaxal system pass through a second fixed point; the two points are extremities of a diameter of a circle orthogonal to the circles of the coaxal system.

d. The square of the distance between two conjugate points equals the sum of the powers of the points with regard to the circle.

142. Theorem (Salmon).   *The distances from the center of a circle to any two points are proportional to the distances from each point to the polar of the other.*

Let O be the center of the circle, A and B any points, A' and B' their inverses, AP and BQ the perpendiculars on the polars $B'P$ and $A'Q$.   If we drop the perpendiculars AA_1 and BB_1 on OB and OA respectively, then we have similar triangles, and

$$\frac{\overline{OA}}{\overline{OB}} = \frac{\overline{OA_1}}{\overline{OB_1}}$$

Again, since A, B, A', B' are concyclic,

$$\frac{\overline{OA}}{\overline{OB}} = \frac{\overline{OB'}}{\overline{OA'}}$$

Combining these proportions,

$$\frac{\overline{OA}}{\overline{OB}} = \frac{\overline{OB'} - \overline{OA}}{\overline{OA'} - \overline{OA}} = \frac{\overline{AP}}{\overline{BQ}}$$

as was to be shown.   From this theorem a number of mildly interesting consequences can be derived.

143. Definitions.   Two triangles are *conjugate* with regard to a circle when the vertices of the one are the respective poles of the sides of the other.   The relation is evidently reciprocal.   A triangle is *self-conjugate* with regard to a circle, if each of its vertices is the pole of the opposite side.

Theorem.   *A triangle self-conjugate with regard to a circle is constructed by taking one vertex arbitrarily, the second on the polar of the first, and the third at the point of intersection of the polars of the first two.*

Theorem. *If a complete quadrangle is inscribed in a circle, the diagonal points, or intersections of opposite connectors, are vertices of a self-conjugate triangle.*

Theorem. *The altitudes of a self-conjugate triangle pass through the center of the circle.*

This follows at once from the definitions.

STEREOGRAPHIC PROJECTION

144. The transformation known as *stereographic projection* is a simple relationship of the points of a plane to those of a sphere, whereby any figure of the plane is transferred to the surface of the sphere.   Because of the direct relation of this transformation to that of inversion, and the consequent preservation of many properties of the figure, we shall outline its essential principles.

Definition.   Given a sphere and a plane tangent to it, the point of the sphere diametrically opposite to the point of contact is taken as center of projection; a point of the sphere and a point of the plane shall be said to correspond by *stereographic projection* if they lie on a line through the center of projection.

We designate the point of contact of sphere and plane as the south pole S of the sphere, the center of projection as the north pole N, the center of the sphere O and its diameter a.

145. Theorems. *Every point of the sphere except the north pole N has a corresponding point in the plane, and to every finite point in the plane corresponds a point of the sphere. Straight lines through S correspond to meridians of the sphere, and circles about S to circles of latitude.   In particular, to the equator of the sphere corresponds a circle in the plane, whose radius is a.*

If we adjoin to the finite plane a single point at infinity, as was suggested in **64**, then this point corresponds to N, and the correspondence is without exception.

Considered as an actual map of the sphere on the plane, the projection furnishes a satisfactory representation of a limited region; in fact, stereographic projection is one of the most frequently used methods of construction of geographic maps.

146. Theorem. *If P and P' are any corresponding points on the sphere and the plane respectively, and u denotes the angle PNS,*

$$\overline{NP} = a \cos u, \quad \overline{NP'} = a \sec u, \quad \overline{SP} = a \sin u, \quad \overline{SP'} = a \tan u$$

Corollary. *If P and Q are symmetrically placed with regard to the equator of the sphere, the projected points P' and Q' in the plane are inverse with regard to the circle S (a) which is the map of the equator.*

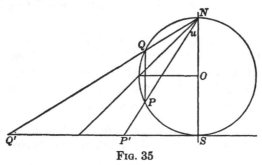

FIG. 35

For if $\angle PNS$ and $\angle QNS$ are complementary, so that $\overline{NP}$ equals $\overline{SQ}$, etc., then

$$\overline{SP'} = a \tan \angle PNS, \quad \overline{SQ'} = a \tan \angle QNS,$$
$$\overline{SP'} \cdot \overline{SQ'} = a^2$$

This theorem furnishes a strikingly interesting interpretation of inversion. To perform an inversion, we first project a figure stereographically on a sphere; then interchange the hemispheres by reflection with regard to the equatorial plane, and project stereographically back into the plane.

Corollary. *Any two corresponding points P, P' are collinear with N, and* $\overline{NP} \cdot \overline{NP'} = a^2$.

147. Theorem. *A stereographic projection is precisely a space inversion with regard to a sphere whose center is N and whose diameter is a.*

We may define and discuss inversion with regard to a sphere by closely imitating the corresponding theory in the plane. Two points are inverse with regard to a sphere O (a) if they are collinear with O, and $\overline{OP} \cdot \overline{OP'} = a^2$. The inverse of a sphere is a sphere, with the special case that a plane corresponds to a sphere through the center of inversion. A sphere orthogonal to the sphere of inversion is unchanged by the transformation. The reader will find it interesting to carry out these analogies in detail.

Now the stereographic projection evidently is just this type of transformation, for $\overline{NP} \cdot \overline{NP'} = a^2$. The given plane is transformed into the given sphere, point by point. From this we may deduce the essential invariant properties of a figure when it is projected on a sphere.

148. Theorem. *A straight line in the plane transforms into a circle through the center of projection, and conversely.*

For the projecting rays lie in a plane through N in either case.

Theorem. *A circle in the plane projects stereographically into a circle of the sphere, and conversely.*

For let a proper circle be given in the plane. Determine the stereographic projection P' of any point P of the circle; and consider the sphere Q passing through the circle and the point P'. Since the sphere passes through two mutually inverse points with regard to the sphere N (a), it is orthogonal to that sphere. It follows that the given circle transforms into a curve lying on the sphere Q, and also on the stereographic sphere. But the intersection of two spheres is a circle. Conversely, this proof is at once reversible.

149. Theorem.   *The angle between two lines in the plane equals the angle between their stereographic projections.*

For the lines AB, AC in the plane transform into circles $NA'B'$, $NA'C'$. The tangents to these circles at N are parallel to AB and AC respectively; and the circles intersect at equal angles at N and A'. Hence the angle at A' equals the given angle at A.

We see then that a stereographic projection transfers a plane figure to a figure on a sphere, and conversely; that angles and circles are preserved, just as in a plane inversion; that a certain inversion in the plane is represented by a simple reflection of the sphere with regard to its equator; and that the transformation itself is a space inversion of the plane into the sphere, with respect to another sphere.

Exercise.   Give complete proofs of the following propositions in this chapter: **116, 118, 120, 122, 127, 128, 133, 135, 136, 137, 140, 141, 143, 145, 146, 147.**

CHAPTER VI
TANGENT CIRCLES

150. In this chapter* we apply the general principles which we have been developing, to the specific problems of systems of tangent circles. To *two* circles there are infinitely many tangent circles, and the first part of the chapter is devoted to the study of these systems of circles, and to a number of interesting configurations associated with them. We then consider various aspects of the problem of Apollonius, to construct a circle tangent to *three* given circles. This famous problem has a finite number of solutions, not exceeding eight. The next problem is that of *four* circles; if four given circles are tangent to a circle, they must satisfy a special condition. The nature of this condition was discovered by J. Casey, and is worked out carefully. The chapter closes with a brief consideration of circles intersecting at constant angles, and at equal angles.

151. Definitions. Two circles are said to be *externally tangent*, when they lie on opposite sides of the tangent at their point of contact; *internally tangent*, when on the same side.

If a circle is tangent to two others, we distinguish between the case that it has *like* contacts of either type, and the case that it is internally tangent to one and externally tangent to the other.

152. Theorem. *Two circles are externally tangent if the distance between their centers equals the sum of their radii; internally tangent, if it equals the difference.*

* As previously indicated, this chapter is not prerequisite for later work and may be omitted without impairing the sequence.

153. Theorem.   *If two tangent circles are subjected to an inversion, the type of tangency is unchanged, except when the center of inversion is inside one circle but outside the other.*

154. Theorem.   *If a circle has like contact with two circles, the points of tangency are collinear with the external homothetic center of the two circles; if unlike contact, with the internal center.   In either case, the common tangents intersect on the radical axis of the two circles.*

This is merely a restatement of **61 e**.

155. We now propose to study in detail the system of circles tangent to two given circles.   Several cases are obviously unlike in detail, according as the given circles do or do not intersect.   In every case, we simplify the figure by inversion.

First, let us assume two fixed circles that do not intersect, and invert them into concentric circles.   Then we recognize two sets of tangent circles, all the circles of each set being equal.   Those of one set lie between the concentric circles and have unlike contact with them, while those of the other set encircle the smaller of the given circles and have internal contact with both.   The circles of the first set are all orthogonal to a circle concentric with the given circles, and at every point of this circle two of them are tangent to each other.   The circles of the second set have no common orthogonal circle, but cut a fixed circle diametrically (**49**).   Each circle of the second set actually intersects all the circles of that set.   We refer to these two sets of circles as the *direct* and the *transverse* system of tangent circles respectively.   From their properties as just described we can by an inversion deduce the properties of the tangent circles to the original circles.

Theorem.   *The circles tangent to two non-intersecting circles consist of two systems, the direct and the transverse system. There is one circle of each system through any chosen point of the region of the plane between the given circles; no circle of the direct series separates the given circles, while every circle* *of the transverse system does so.   If the circles are mutually external, the direct system consists of circles having like contact, and the transverse system of those having unlike contact; if one is within the other, the reverse is true.   There is a common orthogonal circle to the direct system, but not to the transverse.*

156. If the given circles intersect, we may invert them into intersecting straight lines.   The tangent circles then consist of two series, with their centers on the bisectors of the angles formed by the two lines.   The two sets of circles are indistinguishable except that the center of inversion lies in one of the angles, and two of the circles in this angle pass through that point.

Thus in the original figure:

Theorem.   *The circles tangent to two intersecting circles consist of two systems, one of which, called the external series, consists of all circles having like contact, including the tangent lines; the other, the internal, consists of all having unlike contact with the given circles.   Either system has a common orthogonal circle, coaxal with the given circles; and the circles of the system are tangent to one another along this common orthogonal circle.*

By similar reasoning, we see that *the circles touching two given circles which are tangent to each other consist of two series, namely the circles coaxal with them and another series having the properties of the systems just described.*

157. Referring to the inverted figures now and then for inspiration, we are able to make the following assertions about these systems of tangent circles.

Theorem.   *In every case, the circles tangent to two given circles fall into two series, according as they have like or unlike contact with the given circles.   The points of contact with one another of the circles of any series lie on a circle of antisimilitude of the given circles.   If a circle coaxal with the given circles cuts the circles of a tangent series, it cuts them all at a* *constant angle. Any circle orthogonal to the given circles cuts them in four points which are points of tangency of a chain of four tangent circles belonging alternately to the two series (cf. 61 d). The points of contact of any two members of the same series are concyclic. An inversion which exchanges the given circles either leaves unchanged every circle of a tangent series, or interchanges them in pairs. A circle of antisimilitude is orthogonal to all the circles of one series.*

If care is not used in distinguishing the series of tangent circles, difficulties may ensue. For example:

Theorem. *If two circles touch two others, and belong to the same series, the radical axis of either pair passes through the corresponding center of similitude of the other pair.**

The proof is easy.

STEINER CHAINS

158. A Steiner chain of circles is a series of circles, finite in number, each tangent to two fixed circles and to two other circles of the series.

Given two non-intersecting circles, if we start with any circle of their direct tangent system, draw a second circle of the same system tangent to the first, a third tangent to the second, and so on, it is possible but not inevitable that eventually the *n*th circle is tangent to the first. If this occurs, the circles are said to constitute a Steiner chain.

159. Theorem. *A Steiner chain transforms by inversion into a Steiner chain; in particular, any Steiner chain can be inverted into a chain of circles tangent to two concentric circles.*

We therefore derive all the properties of Steiner chains from the properties of the simplified figure, in which the given circles are concentric.

In the three following articles, theorems designated *a* relate

---

* Casey (*Sequel to Euclid*, p. 85) serenely omits to specify that the circles belong to the same tangent system, and his theorem is therefore false.

to the simplified figure, and those designated b to the general figure. Let O be the common center and r, r' the radii of two concentric circles between which there is a Steiner chain; and let C_1, C_2, r_1, r_2, be the centers and the radii of two circles into which these two are inverted.

160. *a.* **Theorem.** *Any set of circles tangent to the given circles, such as a Steiner chain, may be rotated through any angle about O.*

b. **Theorem.** *If two circles admit a Steiner chain, they admit an infinite number, and any one of the direct tangent circles is a member of one chain.*

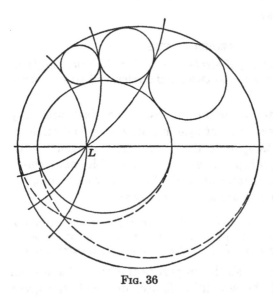

161. *a.* **Theorem.** *The common tangents to the circles of the Steiner chain pass through O, and make equal angles with one another.*

b. **Theorem.** *In the general figure, let K and L be the limiting points of the coaxal system determined by the given circles. (In other words, one of these points is the center of inversion, and the other is the inverse of O.) Then a circle can be drawn through K and L, orthogonal to the given circles, and tangent to any two adjacent circles of the Steiner chain at their point of contact. These circles make equal angles with each other at K and L.*

Fig. 36

162. *a.* **Theorem.** *Two concentric circles admit a Steiner chain of n circles, encircling the common center m times, if the angle subtended at the center by each directly tangent circle is commensurable with* 360°, $\angle TOT' = 360° \cdot m/n$.

b. **Theorem.** *A criterion for two non-concentric circles to admit a Steiner chain is that the angle subtended at K or L by each tangent circle* (as between the orthogonal circles mentioned in **161** *b*) *shall be commensurable with* 360°. *As before, if the angle is* 360° · *m/n, the Steiner chain consists of n circles, overlapping m times.*

163. *a.* **Theorem.** *The angle subtended at O by any directly tangent circle equals the angle at the intersection of two transversely tangent circles whose centers are collinear with O.*

For the radii of the direct and the transverse tangent circles are respectively $\frac{1}{2}(r - r')$ and $\frac{1}{2}(r + r')$, while their distances from O to their centers are respectively $\frac{1}{2}(r + r')$ and $\frac{1}{2}(r - r')$. Congruent triangles are obvious.

b. **Theorem.** *Instead of the angle named in* **162** *b we may take the equal angle between two circles of the transverse tangent series to the two given circles, whose points of contact are on a circle orthogonal to the latter.* (Cf. **157**.)

164. We state without proof the following theorem of Steiner.

Theorem.* *If the number of circles in a Steiner chain is even, any opposite pair of its members touch the given circles at points of a circle orthogonal to the latter. Such a pair of circles are themselves the base circles of another Steiner chain, among whose members are the two given circles; if the characteristic numbers of the two chains* (**162**) *are m, n and m', n', then*

$$\frac{m}{n} + \frac{m'}{n'} = \frac{1}{2}$$

* For proof and references see Coolidge, *Geometry of the Circle*, pp. 31–34.

165. Definition. If A, B, C are three collinear points, the figure bounded by semicircles on AB, BC, CA, all on the same side of the line, is called the *Arbelos* or *Shoemaker's Knife*.

It has some distinctly amusing properties, which were worked out in detail by no less a person than Archimedes.* We content ourselves with a summary of the principal results, leaving proofs for the most part to the reader.

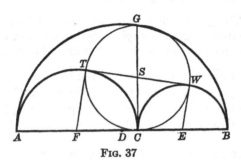

FIG. 37

Let C lie between A and B, and let the perpendicular to ABC at C meet the large circle at G; let the direct common tangent to the circles on AC and BC touch these circles at T and $\overline{W}$ respectively, and intersect CG at S; and denote $\overline{AC}$, $\overline{BC}$, $\overline{AB}$ by $2r_1$, $2r_2$, $2r$.

a. Arc $AGB = $ arc $ATC + $ arc CWB.

b. $\overline{GC}^2 = \overline{TW}^2 = 4r_1r_2$; CG and TW bisect each other at S, which is therefore the center of a circle through C, G, T, W.

c. The area of the arbelos equals the area of the circle having CG and TW as diameters.

d. The lines GA and GB pass respectively through T and W.

* The most complete modern treatment is that of Mackay, *Proceedings Edinburgh Math. Society*, III, 1884, p. 2; other references are given by Simon, *l.c.*, p. 77.

e. If circles are inscribed in each of the curvilinear triangles ACG, BCG, these circles are equal, the diameter of either being

$$\frac{\overline{AC} \cdot \overline{BC}}{\overline{AB}} = \frac{r_1 r_2}{r}$$

Let the first circle touch CG, and arcs AC and AB, at L, M, N respectively, and let QL be a diameter. Then because M and N are centers of similitude, AL and CQ meet at M, and AQ and BL at N. Extend AN to meet CG at Y, then the altitudes of triangle ABY meet at L; BY and QC are parallel, being perpendicular to AL. Then

$$\frac{\overline{QL}}{\overline{AC}} = \frac{\overline{QY}}{\overline{AY}} = \frac{\overline{CB}}{\overline{AB}}$$

whence the result.

f. The common tangent to the two circles at M passes through B. (For L, N, A, C are concyclic.)

g. It can be shown by a long computation that *the smallest circle that is tangent to and circumscribes the two circles of (e) is equal to the circle on CG, and therefore equal in area to the arbelos.*

h. **Theorem of Pappus.** *In the arbelos let us consider a chain of circles c_1, c_2 . . . all tangent to the circles on AB and AC; c_1 shall be tangent to the circle on BC, c_2 to c_1, and so on. Then if r_n represents the radius of c_n, h_n the distance from its center to ACB,*

$$h_n = 2\, n\, r_n$$

The proof depends on an inversion with A as center. The series of circles are carried into equal circles tangent to two parallel lines; and the result is based on proportions.

THE PROBLEM OF APOLLONIUS

166. We come next to one of the famous problems of geometry, associated with the name of Apollonius, namely:

To construct the circle or circles tangent to three given circles.

It will be assumed that the given circles are not coaxal. (We observe that for three coaxal circles there is no solution except when the coaxal system consists of tangent circles.) The contact of the required circle with each of the given circles may be external or internal; hence we anticipate that there will in general be eight circles satisfying the conditions of the problem. Consideration of the various possibilities shows us that in some cases there will actually be eight solutions, while in other cases there are none.

We first analyze the problems by methods similar to those used by Apollonius, but making use of more modern terminology; then we shall consider a simpler modern solution.

167. We first note that if two circles are internally tangent, and their radii are both increased or decreased by the same amount, the resulting circles are still tangent. Likewise, if two externally tangent circles are modified by adding to one radius the amount subtracted from the other, the new circles are tangent. Consequently, if the radius of a circle is changed, while the radii of all circles tangent to it are increased or decreased correspondingly, the resulting circles are tangent.

Theorem. *The problem of Apollonius is equivalent to the problem of constructing a circle through a given point and tangent to two given circles.*

For let it be required to construct a circle tangent in assigned ways to three given circles; say, for definiteness, externally tangent to all three. If such a circle exists, let its radius be increased by the radius of the smallest given circle, while the radii of the three given circles are simultaneously decreased by the same amount. Then the required circle passes through a known point and touches two known auxiliary circles.

168. Theorem. *A circle having assigned contacts with two* *given circles and passing through a given point passes also through a second point. Hence the problem of constructing such a circle is equivalent to that of constructing a circle through two given points and tangent to a fixed circle.*

For let the required circle be tangent to the given circles at S and T; let A be the given point. We know (**61** *e*, **154**) that the line ST passes through a center of similitude of the given circles C; and if CA meets the required circle again at A', then

$$\overline{CA} \cdot \overline{CA'} = \overline{CS} \cdot \overline{CT}$$

Therefore A' is a fixed point which can easily be determined. If the product is positive, there is a circle of antisimilitude of the given circles with center at C, and A' is the inverse of A with regard to this circle.

169. Problem. *To construct a circle through two given points, and tangent to a given circle.* (Cf. **56**.)

Let A, A' be the given points, c the given circle. Draw any circle through A, A' and meeting c at P, Q; and let PQ meet AA' at O. Then the power of O with regard to the circle equals $\overline{OA} \cdot \overline{OA'}$. Drawing tangents to c from O, the points of tangency are also points of tangency for the required circles. We have thus reduced the problem to that of drawing a circle through three known points.

170. The foregoing solution yields two circles, and therefore two solutions of the original problem, which are paired by virtue of having unlike contacts with each of the three given circles. We obtain the four pairs of solutions by the four different possibilities in **167**; according as we increase the radii of the two larger circles by the amount of the radius of the smallest, diminish both, or increase one and diminish the other.

171. For extensive references to the modern elaborations of this and related problems, one may consult the bibliography of Simon.* Of the numerous solutions, the best known and neatest is the following, due to Gergonne. This construction, like the more elementary one already considered, yields the required circles in pairs, the members of each pair having unlike contacts with the given circles.

Construction. *Determine the six homothetic centers of similitude of the given circles; these lie three by three on four lines. We determine the poles of one of these lines with regard to the three circles; and connect these poles with the radical center of the circles. If these connectors meet the respective circles, the three pairs of intersections are the points of tangency of two of the required circles.*

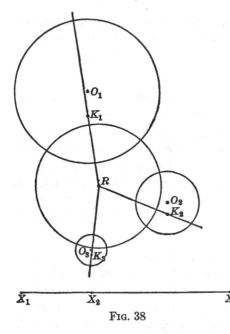

Let the given circles be c_1, c_2, c_3, their centers O_1, O_2, O_3. Let us consider a pair of the required tangent circles, say those which have like contacts with the three given circles. Designating these by c and $\bar{c}$, the points of tangency by P_1, P_2, P_3, Q_1, Q_2, Q_3 respectively, and the external centers of similitude of the given circles by X_1, X_2, X_3 respectively, we recall first that P_2P_3 and Q_2Q_3 pass through X_1, and that P_2, P_3 and Q_2, Q_3 are pairs of antihomologous points;

FIG. 38

* *L.c.*, pp. 97–105.

$$\overline{X_1P_2}\cdot\overline{X_1P_3} = \overline{X_1Q_2}\cdot\overline{X_1Q_3}$$

It follows that X_1 lies on the radical axis of c and $\bar{c}$; and similarly for X_2 and X_3. Thus these three centers of similitude are collinear.

Next, we observe that the given circle c_1 has unlike contacts with c and $\bar{c}$; hence the line joining the points of contact P_1Q_1 passes through the internal center of similitude of c and $\bar{c}$.

But this same point is the radical center of the three given circles. For an inversion with regard to their common orthogonal circle (or if necessary an inversion followed by a rotation through 180°) leaving the three given circles in place, exchanges the circles c and $\bar{c}$. Hence P_1Q_1, P_2Q_2, P_3Q_3 pass through the radical center R.

Finally, the tangents to c_1 at P_1 and Q_1 meet on the radical axis $X_1X_2X_3$; that is, the pole of P_1Q_1 is on $X_1X_2X_3$. It follows that the pole of $X_1X_2X_3$ with regard to c_1 is on the line P_1Q_1.

Hence the construction: to determine P_1Q_1, we first determine the line $X_1X_2X_3$, and its pole with regard to c_1; and also the radical center R. The line from R to the pole cuts c_1 in the points P_1 and Q_1.

By similar argument we find that each external center of similitude is collinear with two internal centers; the poles of the three lines thus determined are connected with the same radical center R, to find the points of contact of the other three pairs of tangent circles. If it happens that a set of lines from R fail to meet the respective circles, the corresponding tangent circles do not exist.

CASEY'S THEOREM

172. Theorem. *Four circles c_1, c_2, c_3, c_4 are tangent to a circle or straight line, if and only if*

$$t_{12}t_{34} \pm t_{13}t_{42} \pm t_{14}t_{23} = 0$$

where t_{12}, for instance, is a common tangent to c_1 and c_2.

The remarkable theorem here loosely stated was first given by John Casey,[*] but in incomplete form; for he establishes only that when four circles touch a circle, this equation is true. The converse theorem, which is the more important in the applications, has frequently been proved under various restrictions.[†] We notice that the theorem may be regarded as an elaboration of the theorem of Ptolemy. (Cf. **92, 117.**)

173. Before proving the theorem of Casey, we establish the following lemmas.

Theorem. *The quotient of the square of the length of the direct common tangent to two circles, divided by the product of the radii, is unchanged when the circles are subjected to an inversion, provided the center of inversion lies inside both circles, or outside both. The same is true of the transverse common tangent; in other words, if r_1, r_2 are the radii of two circles, and t_{12} and $\bar{t}_{12}$ their direct and transverse common tangents respectively, each of the quantities $\dfrac{t_{12}^{\;2}}{r_1 r_2}$, $\dfrac{\bar{t}_{12}^{\;2}}{r_1 r_2}$ is invariant with regard to an inversion whose center is inside both circles or outside both.*[§]

[*] *Sequel to Euclid*, p. 102.

[†] See, for instance, Lachlan, *l.c.*, pp. 244–51. We shall follow with some modifications a method of proof ascribed by Lachlan to H. F. Baker.

[§] It must be noted that if two circles are subjected to an inversion whose center is inside one and outside the other, the above theorem is not valid. If the two circles, and therefore the inverse pair, do not intersect, then one pair has four common tangents, and the other pair has none, so that the theorem cannot even be stated. On the other hand, if the given circles intersect, each pair has direct common tangents, but the theorem is not true of these tangents. Casey (*l.c.*, VI, 9) does not notice this failing case, and erroneously states without qualification that the fraction $\dfrac{t_{12}^{\;2}}{r_1 r_2}$ is unchanged by an inversion. It is possible to circumvent the difficulties that arise in this situation, by setting up the formulas for the squares of common tangents

$$t_{12}^{\;2} = \overline{C_1 C}_2^{\;2} - (r_1 - r_2)^2, \quad \bar{t}_{12}^{\;2} = \overline{C_1 C}_2^{\;2} - (r_1 + r_2)^2$$

and choosing names for these which will still be admissible when they are negative and the common tangents therefore non-existent. Then, in the case which causes trouble, these two expressions are interchanged. The theorems as stated in the text, however, are sufficient for our purpose.

Let the given circles be $C_1(r_1)$, $C_2(r_2)$, and let C_1C_2 cut the circles at P_1, Q_1, P_2, Q_2, so that P_1Q_1 and P_2Q_2 are in the same direction as C_1C_2. Then if $d = \overline{C_1C_2}$,

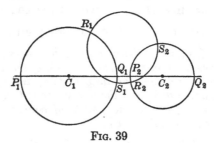

FIG. 39

$$\frac{\overline{P_1Q_2}\cdot\overline{Q_1P_2}}{\overline{P_1Q_1}\cdot\overline{P_2Q_2}} = \frac{(d+r_1+r_2)\,(d-r_1-r_2)}{2\,r_1\cdot 2\,r_2} = \frac{d^2-(r_1+r_2)^2}{4\,r_1r_2}$$

$$\frac{\overline{P_1P_2}\cdot\overline{Q_1Q_2}}{\overline{P_1Q_1}\cdot\overline{P_2Q_2}} = \frac{(d+r_1-r_2)\,(d-r_1+r_2)}{2\,r_1\cdot 2\,r_2} = \frac{d^2-(r_1-r_2)^2}{4\,r_1r_2}$$

These numerators, if positive, represent respectively the square of the transverse and of the direct common tangent.

Next, if any circle orthogonal to the given circles cuts them (in the same order as P_1, Q_1, P_2, Q_2) at R_1, S_1, R_2, S_2, then

$$\frac{\overline{R_1S_2}\cdot\overline{S_1R_2}}{\overline{R_1S_1}\cdot\overline{R_2S_2}} = \frac{\overline{P_1Q_2}\cdot\overline{Q_1P_2}}{\overline{P_1Q_1}\cdot\overline{P_2Q_2}}, \qquad \frac{\overline{R_1R_2}\cdot\overline{S_1S_2}}{\overline{R_1S_1}\cdot\overline{R_2S_2}} = \frac{\overline{P_1P_2}\cdot\overline{Q_1Q_2}}{\overline{P_1Q_1}\cdot\overline{P_2Q_2}}$$

For by an inversion we can exchange the circle $R_1S_1R_2S_2$ with the line $P_1Q_1P_2Q_2$; and these fractions are invariant (**68** c).

Now let the given circles be transformed by inversion into $C_1'(r_1')$, $C_2'(r_2')$; and let R_1, S_1, R_2, S_2 be carried into R_1', S_1', R_2', S_2'. Then

$$\frac{\bar{t}_{12}{}^2}{4\,r_1r_2} = \frac{\overline{R_1S_2}\cdot\overline{S_1R_2}}{\overline{R_1S_1}\cdot\overline{R_2S_2}} = \frac{\overline{R_1'S_2'}\cdot\overline{S_1'R_2'}}{\overline{R_1'S_1'}\cdot\overline{R_2'S_2'}} = \frac{\bar{t}_{12}'{}^2}{4\,r_1'r_2'}$$

and similarly for t_{12}. (It can be shown that if the center of inversion is inside one circle and outside the other, the order of R_1', S_1', R_2', S_2' is not the same as that of R_1, S_1, R_2, S_2; so that t_{12} and $\bar{t}_{12}$ are interchanged, with some changes in sign.)

174. We proceed to a proof of the Casey criterion.

Theorem.* *Let four circles c_1, c_2, c_3, c_4 be tangent to a circle k; and either none of them surround k, or all of them do so. Let T_{12} denote the direct common tangent to c_1 and c_2 when these have like contact with k, and the transverse if they have unlike contact with k. Then if the points of contact of c_1 and c_4 separate those of c_2 and c_3 on k,*

$$T_{13}T_{34} + T_{13}T_{24} - T_{14}T_{23} = 0$$

First, we take a center of inversion on the circle k, thereby transforming it into a straight line and the four circles c_1, c_2, c_3, c_4, into circles tangent to this line. Since the center of inversion is inside all four circles or outside all four, the type of contact is in no case changed. If the points of contact on this line are A_1, A_2, A_3, A_4, we have at once

$$\overline{A_1A_2}\cdot\overline{A_3A_4} + \overline{A_1A_3}\cdot\overline{A_2A_4} - \overline{A_1A_4}\cdot\overline{A_2A_3} = 0$$

But $\overline{A_1A_2}$ is the common tangent T_{12}, in the inverted figure. We therefore divide each member of this equation by the product of the square roots of the radii of the circles, apply **173**, and clear again of fractions, whence the equation as stated in the theorem.

175. Exercises. *a. The invariants of* **173**, *when the given circles intersect, represent respectively the square of the sine and the square of the cosine of half their angle of intersection.*

(Invert the circles into intersecting lines.)

---

* It does not appear that most of the writers have stated the exact limitations on the validity of this theorem. Casey, Lachlan, and others, do not sufficiently restrict it; while Coolidge unnecessarily limits its scope by stipulating that the circles be mutually external. It will be evident that the theorem as here given includes all the cases where the tangents in question exist; in other words, all the cases in which the formula can be stated in terms of real numbers.

b. When four given circles are tangent to a null-circle their common tangents satisfy the Casey equation.

(By inverting the circles into straight lines the Casey equation is transformed into a trigonometric identity.)

176. When four circles are tangent to a circle, as in the foregoing theorem, there are three possible cases, according as all the circles are on the same side of k, three on one side and one on the other, or two on each side; and accordingly, all six of the tangents in the Casey equation, or the three which touch three of the circles, or one pair only, will be direct tangents, while the rest will be transverse. We incorporate this conclusion into the statement of the converse theorem:

Theorem. *If certain common tangents to four circles c_1, c_2, c_3, c_4 satisfy an equation of the form*

$$T_{12}T_{34} \pm T_{13}T_{24} \pm T_{14}T_{23} = 0$$

then these circles are tangent to a circle k, as follows:

(a) if all the T's are direct common tangents, then k has like contact with all the circles.

(b) if the T's from one circle are transverse, while the other three are direct, then this one circle has contact with k unlike that of the other three.

(c) if the given circles can be so paired that the common tangents to the circles of each pair are direct, while the other four are transverse, then the members of each pair have like contact with k.

The proof involves several steps. We begin by diminishing the radius of the smallest circle, say c_4, to zero, while simultaneously increasing or decreasing each of the other radii by the same amount, so that all six of the common tangents are unchanged in length and direction. Under each of the hypotheses (a), (b), (c) this can be done, and the point C_4, which replaces the fourth circle, will be exterior to the other three transformed circles.

We now perform an inversion with C_4 as center and any convenient radius R. In accordance with **173**, we have

$$T_{12} = T_{12}' \sqrt{\frac{r_1 r_2}{r_1' r_2'}} \text{ etc.}$$

Also, by **71**, $\quad \dfrac{r_3'}{r_3} = \dfrac{R^2}{T_{34}^2} \quad$ whence $\quad T_{34} = R \sqrt{\dfrac{r_3}{r_3'}}$

Substituting in the given equation, we have, after cancellation,

$$T_{12}' + T_{23}' + T_{31}' = 0$$

where either all three are direct tangents, or one is direct and the other two are transverse. If we now again diminish the smallest of the three circles to a point C_3', changing the other radii simultaneously, we have only to show that C_3' is on the common tangent to the surviving circles, c_1'' and c_2''. Let PQ be a common tangent to these circles whose length is T_{12}'; and lay off $\overline{PS}$ equal to T_{13}' so that by the equation, $\overline{QS}$ equals T_{23}'. The locus of a point from which the tangent to either circle has the given value is a circle concentric with it; and these two circles intersect once on each common tangent. Therefore either S or the symmetrical point on the other tangent is the point C_3'; therefore c_1', c_2', c_3' are tangent to a line; therefore the original circles are tangent to a circle.

177. An alternative form of the criterion is based on the interpretation suggested in **175**. We content ourselves with suggesting the possibilities:

Theorem. *If four circles c_1, c_2, c_3, c_4, intersecting at angles ω_{12}, etc., have like contacts with a circle k, then*

$$\sin \frac{\omega_{12}}{2} \sin \frac{\omega_{34}}{2} \pm \sin \frac{\omega_{13}}{2} \sin \frac{\omega_{24}}{2} \pm \sin \frac{\omega_{14}}{2} \sin \frac{\omega_{23}}{2} = 0$$

When some of the contacts are unlike, the corresponding terms are cosines; and conversely, if such an equation holds the circles are tangent to a fifth circle.

178. We shall not elaborate the possible applications of this theorem; one may consult Lachlan (*l.c.*), where will be found a more complete treatment of all these problems, together with references to the original memoirs of Larmor and others. We suggest a few immediate corollaries.

a. **Theorem.** *Two pairs of circles, mutually inverse with regard to a circle, have four common tangent circles.*

b. **Theorem.** *The four circles which are tangent to three given non-concurrent lines are touched by a circle.*

In this case we can easily determine the lengths of the common tangents in terms of the sides of the triangle, and show that the Casey equation is satisfied. Moreover, by application of **117** and using the mid-points of the sides of this triangle, we may show that the circle through the mid-points of the sides is tangent to the four circles in question. This famous theorem is the subject of Chapter XI, and may be studied in detail there.

c. **Theorem of Hart.** *The circles tangent to three given circles (cf.* **166**) *have the property that certain fours of them are tangent to still other circles. Specifically, any one of the eight circles, and the three others that have like contact with two of the given circles and unlike with the third, are tangent to a circle.*

We shall follow Casey's proof of this theorem, which evidently does not pretend to be entirely adequate nor to cover all cases.

Let the given circles be a_1, a_2, a_3; let c_4 be any tangent circle, and c_1, c_2, c_3 any three others, so that c_1 and c_4, for instance, have unlike contacts with a_1 and like contacts with a_2 and a_3. Then denoting direct tangents by t, transverse by t, since all four circles touch a_1,

$$\bar{t}_{12}t_{34} = \bar{t}_{13}t_{24} + \bar{t}_{14}t_{23}$$

and a_2, $\bar{t}_{12}t_{34} = \bar{t}_{13}\bar{t}_{24} + t_{14}\bar{t}_{23}$

and a_3, $t_{14}\bar{t}_{23} = t_{12}\bar{t}_{34} + \bar{t}_{13}t_{24}$

Combining these equations, the result is

$$t_{12}\bar{t}_{34} = \bar{t}_{13}t_{24} + \bar{t}_{14}t_{23}$$

which establishes the existence of a circle having like contacts with c_1, c_2, c_3, and the opposite contact with c_4.

Thus each of the eight tangent circles determines a new circle, called a Hart circle, which touches it and has unlike contact with three others; in other words, the eight circles are tangent to the three given circles, and, also, four by four to eight others, the Hart circles.

We may add that they are also tangent four by four to six other circles, of a different type.   For we recall that the eight are paired by virtue of being mutually inverse with regard to the common orthogonal circle of a_1, a_2, a_3; any circle orthogonal to the latter, and tangent to two of the eight which are not paired, will be tangent also to their inverses.

CIRCLES INTERSECTING AT GIVEN ANGLES

179. Our study of orthogonal circles and tangent circles suggests the more general problems of circles intersecting at given angles or at equal angles.   We shall very briefly consider some of the more obvious possibilities.

In order to make our statements definite, we specify as the angle between two intersecting circles the angle between their radii at either point of intersection.   The angle is signless and is determined by its cosine, namely:

$$\cos\theta = \frac{r_1{}^2 + r_2{}^2 - d^2}{2\,r_1 r_2}$$

where r_1 and r_2 are the radii of the circles and d is the distance between their centers.

180. Theorem. *Given two non-concentric circles $C_1(r_1)$ and $C_2(r_2)$, where $\overline{C_1C_2} = d$; let any circle $C(r)$ cut these at angles θ_1 and θ_2 respectively, and let h be the perpendicular from C on the radical axis of the given circles.   Then*

$$r\,(r_1 \cos \theta_1 - r_2 \cos \theta_2) = d \cdot h$$

For $\quad \cos \theta_1 = \dfrac{r_1^{\,2} + r^2 - \overline{C_1C}^2}{2\,r_1 r}; \quad \cos \theta_2 = \dfrac{r_2^{\,2} + r^2 - \overline{C_2C}^2}{2\,r_2 r}$

Clearing of fractions, subtracting, and applying **113**, we have the result.

Exercise. Extract various theorems from the formula just given, including some of the results below; also results concerning orthogonal circles.

If the given circles are concentric, the foregoing theorem is meaningless, since the radical axis is at infinity.   We have the following substitute theorem, which may be proved as an exercise:

Theorem. *If a circle of radius r cuts two concentric circles whose radii are r_1 and r_2, at angles θ_1 and θ_2 respectively, then*

$$r(r_1 \cos \theta_1 - r_2 \cos \theta_2) = \tfrac{1}{2}(r_1^{\,2} - r_2^{\,2})$$

181. In order to consider the properties of those circles which cut two given circles at constant angles, we may depend on the formula of **180**.   It is more enlightening, however, to simplify the figure by inversion, transforming two intersecting circles into straight lines and two non-intersecting circles into concentric circles.   We can set down at once the following results.

Theorem. *The system of circles cutting two fixed circles at given angles are cut at a fixed angle by any circle coaxal with the fixed circles which meets them at all.*

Theorem. *The circles cutting two intersecting circles at given angles have a common orthogonal circle.   The circles* *cutting two non-intersecting circles at constant angles have either a common orthogonal circle or a circle cutting them all diametrically. In either case, the circle in question is coaxal with the given circles.*

Theorem. *In general, the circles which cut two given circles at given angles are tangent to two fixed circles coaxal with the given circles; but these two circles may be non-existent (imaginary).*

182. We may discuss the problem of constructing a circle which cuts three given circles at given angles. Disregarding all considerations of real and imaginary, we see that the required circles are orthogonal to three fixed circles, coaxal with respective pairs of given circles. They are also tangent to three pairs of fixed circles of the same coaxal systems. We see intuitively that there will generally be two solutions, representing circles cutting the given circles at the given angles or at the supplements of all three, and the solution can be made to depend on that of the problem of Apollonius.

Another problem is that of circles which cut given circles at equal but unspecified angles. Just as three given circles have in general one common orthogonal circle, so in the light of the foregoing there is one circle cutting three given circles at angles equal to a given angle, and another cutting them at the supplementary angle. It follows that there is in general one circle cutting four given circles at equal angles.*

Exercise. In this chapter theorems are left unproved, and proofs are to be completed by the reader, in the following sections; **152–157, 159–163, 165, 175, 178, 180, 181.**

---

* For fuller treatment of these topics, one may consult Lachlan, chap. XV; Coolidge, chaps. I, II, III.

CHAPTER VII

THE THEOREM OF MIQUEL

183. We now begin the systematic study of the triangle, and of the numerous points, lines, and circles associated with it. Except for a very few theorems known to the ancients, the development of this subject has taken place almost entirely in the nineteenth and twentieth centuries. We shall attempt a broad outline of the central and most important theorems, with fairly numerous applications; but we can by no means hope to exhaust the field, which has been the subject of hundreds of researches and published papers.

In the present chapter, the central theorem is a remarkably simple one, whose significance seems not to have been fully appreciated. We shall find it to be the source of many theorems of wide applicability.

184. Theorem. *If a point is marked on each side of a triangle, and through each vertex of the triangle and the marked points on the adjacent sides a circle is drawn, these three circles meet at a point.*

It is, of course, understood that the marked points may lie on the extensions of the sides. If in particular one of them is at a vertex of the triangle, the circle through two coincident points is drawn tangent to the line on which they lie.

Let the triangle be $A_1A_2A_3$; the marked points P_1, P_2, P_3 on A_2A_3, A_3A_1, A_1A_2 respectively. We are to prove that the circles $A_1P_2P_3$, $A_2P_3P_1$, $A_3P_1P_2$ meet at a point. Since this theorem furnishes an excellent illustration of the advantages of the notation of directed angles, we shall give the proof in two forms.

For a first proof, let us take the marked points on the sides of the triangle rather than their extensions, as shown in the figure.   Also let the point of intersection P of two of the

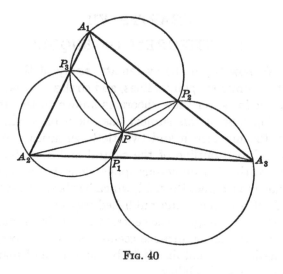

Fig. 40

circles, $A_1P_2P_3$ and $A_2P_3P_1$ lie within the triangle.   Then at once

$$\angle P_2PP_3 = 180° - \alpha_1$$
$$\angle P_3PP_1 = 180° - \alpha_2$$

Combining these, we easily see that

$$\angle P_1PP_2 = 180° - \alpha_3$$

which shows that P, P_1, P_2, A_3 are concyclic.

Now this, which is the standard proof, is obviously inadequate.   For we cannot assume, in general, that P will fall inside the triangle, and the angles considered in the proof may therefore be either supplementary or equal; so that a complete proof will necessitate the consideration of many different cases.   This difficulty is obviated by adopting the convention of directed angles; with that method, a single proof covers all possible cases.

For as before, let P denote the point common to the circles $A_1P_2P_3$, $A_2P_3P_1$, then

$$\measuredangle\, PP_2A_1 = \measuredangle\, PP_3A_1, \text{ and } \measuredangle\, PP_3A_2 = \measuredangle\, PP_1A_2 \quad (19)$$

That is to say,

$$\measuredangle\, PP_2, a_2 = \measuredangle\, PP_3, a_3 \text{ and } \measuredangle\, PP_3, a_3 = \measuredangle\, PP_1, a_1$$

Hence $\measuredangle\, PP_2, a_2 = \measuredangle\, PP_1, a_1,$

That is, $\measuredangle\, PP_2A_3 = \measuredangle\, PP_1A_3$

which proves (19) that P, P_1, P_2, A_3 are on a circle.

The source of this theorem is doubtful.   It was explicitly stated and proved by A. Miquel in 1838, though its truth was probably recognized much earlier.   It has not received from most writers the attention that it deserves, but we shall make it the cornerstone of our geometric structure.   For the sake of definiteness, the theorem will be called the Miquel theorem, the point P will be called the Miquel point for the triad $P_1P_2P_3$ with respect to triangle $A_1A_2A_3$, $P_1P_2P_3$ is a Miquel triangle of P, and the circles are Miquel circles.

185. Theorem.   *The lines from the Miquel point to the marked points make equal angles with the respective sides.*

This is a by-product of the proof of the main theorem.

186. Theorem.   *The angles of the figure satisfy the equation*

$$\measuredangle\, A_2PA_3 = \measuredangle\, A_2A_1A_3 + \measuredangle\, P_2P_1P_3$$

For

$$\measuredangle\, A_2PA_3 = \measuredangle\, A_2PP_1 + \measuredangle\, P_1PA_3 = \measuredangle\, A_2P_3P_1 + \measuredangle\, P_1P_2A_3$$

But

$$\measuredangle\, A_2P_3P_1 + \measuredangle\, P_1P_2A_3 = \measuredangle\, A_1A_2, P_3P_1 + \measuredangle\, P_1P_2, A_1A_3$$

$$= \measuredangle\, A_1A_2,\, A_1A_3 + \measuredangle\, P_1P_2,\, P_1P_3$$
$$= \measuredangle\, A_2A_1A_3 + \measuredangle\, P_2P_1P_3$$

as was to be shown. This formula is fully as important and as useful to us as the main theorem; though not given by Miquel we shall call it the Miquel equation because of its close association with that theorem. Its resemblance to the fundamental angle theorem of inversion (**75**) will suggest to the reader some possible theorems; these will be developed later.

187. Theorem. *Conversely, if P is a fixed point in the plane of triangle $A_1A_2A_3$, it is possible to determine in an infinite number of ways a Miquel triangle for P.*

For we may start by drawing from P any set of lines making equal angles with the sides; or by passing any circle through P and one of the vertices.

If from any point P we draw three lines which make equal angles with the sides of a triangle, and which may be thought of as a rigid system rotating about P, their intersections with the corresponding sides trace out all the possible Miquel triangles of P.

188. Theorem. *All the Miquel triangles of a given point P are directly similar, and P is the center of similitude or self-homologous point in every case (**33**).*

For if $P_1P_2P_3$ is any Miquel triangle of a fixed point P, it follows at once from **186** that the angles of triangle $P_1P_2P_3$ are fixed in magnitude and direction also

$$\measuredangle\, P_2P_3P = \measuredangle\, P_2A_1P = \measuredangle\, A_3A_1P$$

showing that P is self-homologous.

Corollaries:

a. The centers of any set of Miquel circles are vertices of a triangle similar to the given triangle.

b. If two or more directly similar triangles are drawn with homologous vertices on the respective sides of a given triangle, they have the same Miquel point, which is their center of similitude.

c. If corresponding vertices of several directly similar triangles are collinear, they have a common center of similitude.

d. If three circles are concurrent at a point, it is possible to start at any point of one of the circumferences, and draw a triangle whose vertices lie on the circles and whose sides pass through the corresponding intersections; and all triangles thus drawn are similar.

This may be proved directly, or by inversion; if we apply an inversion to the original theorem (**184**), the result is the theorem before us.

If we take any of the seven points A_1, A_2, A_3, P_1, P_2, P_3, P of the original figure as center of inversion, we derive another figure like the original. If, however, we take a center of inversion not related to the figure, we obtain a generalization of the theorem whereby the sides of the original triangle are replaced by arcs of circles passing through a fixed point. In other words, the theorem of Miquel is equivalent to the following:

e. If the circles $A_1A_2B_3$, $A_2A_3B_1$, $A_3A_1B_2$ are concurrent at a point O, the circles $A_1B_2B_3$, $A_2B_3B_1$, $A_3B_1B_2$ are concurrent at a point P.

THE PEDAL TRIANGLE AND CIRCLE

189. Definition.   The *pedal triangle* of a point with regard to a triangle is that triangle whose vertices are the feet of the perpendiculars from the point to the sides of the given triangle. The circle circumscribed about the pedal triangle is called the *pedal circle*.

Obviously the pedal triangle is one of the Miquel triangles of a point; and it need hardly be added that it is the most important one. Evidently the form of the pedal triangle of a given point is determined by **186**.

190. Theorem.   *The sides of the pedal triangle of a point P are:*

$$\overline{P_2P_3} = \overline{A_1P}\ sin\ \alpha_1 = \frac{\overline{A_1P}\cdot a_1}{2\,R}, \text{ etc.}$$

For $\overline{P_2P_3}$ is a chord of the circle on A_1P as diameter; and in that circle the angle inscribed in arc P_2P_3 equals α_1.

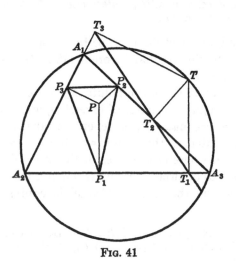

FIG. 41

Corollaries:

a. The sides of any Miquel triangle of a point are proportional to the products of the corresponding sides of the given triangle and the distances from the opposite vertices to the given point.

b. The only point whose Miquel triangles are similar to the given triangle in the sense $P_1P_2P_3 \sim A_1A_2A_3$ is equidistant from the vertices of the triangle, and therefore is the circumcenter O.

c. If the form of a triangle to be inscribed in a given triangle is given, the position of the Miquel point is determined either

by **186** *or by* **190** *a.   In the latter case it may be either of
two points, as seen in* **95***; the Miquel triangles of these two
points are inversely similar.   (Cf.* **201***.)*

SIMSON LINE

191. A case of special interest arises when the pedal
triangle of a point reduces to a straight line; in other words,
when the feet of the perpendiculars from a point are collinear.   We have at once the following theorem:

Theorem. *The Miquel point of any set of collinear points
on the sides of a triangle is on the circumscribed circle; and
conversely, any Miquel triad for a point on the circumscribed
circle are collinear.*

For in the equation of **186**, if $\measuredangle\ T_2T_1T_3 = 0$, then

$$\measuredangle\ A_2TA_3 = \measuredangle\ A_2A_1A_3$$

and conversely.   In particular (figure 41):

192. Theorem. *The feet of the perpendiculars to the sides
of a triangle from a point are collinear, if and only if the
point is on the circumscribed circle of the triangle.*

Definition. The line through the feet of the perpendiculars to the sides of a triangle from a point of its circumcircle
is called the *pedal line,* or *Simson line,* of the point with regard to the triangle.

Historical. In the nineteenth century it was generally
supposed that this theorem was due to Robert Simson (1687–
1768), and his name was attached to the line.   It has been
shown, however, by that diligent investigator J. S. Mackay *
that the theorem is not to be found in any of Simson's writ-
ings, nor is there any evidence that it was known to him.
Mackay finds that the error originated in a careless statement
of the French geometer Servois, who referred to "le théorème

* *Proceedings of Edinburgh Math. Society,* IX, 1890, pp. 83–91; see also Muir,
ibid., III, 1884, p. 104.

suivant, qui est, je crois, de Simson." Later, in his treatise on projective geometry, Poncelet reproduced the ascription without the qualifying phrase, and thus perpetuated the error. The theorem was first discovered in 1797 by one William Wallace; its history is given in detail in the paper of Mackay. Following Mackay's example, some geometers have discarded the familiar term "Simson line," and designated the line as the "Wallace line"; undoubtedly the noncommittal "pedal line" would be in many ways preferable, but we shall adhere to the traditional term. Many theorems about these lines will be established first and last; a few obvious properties will be suggested at this time.

The theorem of Ptolemy can be proved at once as a corollary, using the formula of **190**.

193. Theorem. *The Simson line of any vertex is the altitude through that vertex; that of the point diametrically opposite to a vertex is the corresponding side.*

194. Theorem. *If $T_1 T_2 T_3$ is the Simson line of a point T of the circumscribed circle, then triangles $T T_1 T_2$ and $T A_2 A_1$ are directly similar.*

For the angles at T are equal, and the including sides are proportional:

$$\angle\, T_1 T T_2 = \angle\, A_2 A_3 A_1 = \angle\, A_2 T A_1$$

$$\frac{\overline{TT_1}}{\overline{TT_2}} = \frac{\sin \angle\, A_2 A_3 T}{\sin \angle\, A_1 A_3 T} = \frac{\sin \angle\, A_2 A_1 T}{\sin \angle\, A_1 A_2 T} = \frac{\overline{A_2 T}}{\overline{A_1 T}}$$

Corollaries:

a. $$\overline{TA_1} \cdot \overline{TT_1} = \overline{TA_2} \cdot \overline{TT_2} = \overline{TA_3} \cdot \overline{TT_3}$$

b. $$\frac{\overline{TT_1} \cdot \overline{T_2 T_3}}{a_1} = \frac{\overline{TT_2} \cdot \overline{T_3 T_1}}{a_2} = \frac{\overline{TT_3} \cdot \overline{T_1 T_2}}{a_3}$$

c. $$\frac{a_1}{\overline{TT_1}} + \frac{a_2}{\overline{TT_2}} + \frac{a_3}{\overline{TT_3}} = 0$$

195. Theorem. *The projection of any side of a triangle on the Simson line of a point equals the distance between the feet of the perpendiculars from the point to the other two sides.*

For the projection of A_2A_3 on the line $T_1T_2T_3$ is

$$\overline{A_2A_3} \cos \angle A_3T_1T_3 = a_1 \sin \angle TA_2T_3 = a_1 \sin \angle TA_3A_1$$

which by **190** is also precisely the length of $\overline{T_2T_3}$.

196. The general theorem, that when three collinear points are marked on the sides of a triangle their Miquel point is on the circumcircle of the triangle, may be expressed in the following striking form.

Theorem. *The circumscribed circles of the four triangles formed by four lines in general position are concurrent.*

For if we fix the attention on one of the triangles, the fourth line marks a point on each of its sides, and the Miquel circles are merely the circumcircles of the other three triangles. Since the marked points are collinear, these circles are concurrent at a point of the first circle. Again:

197. Theorem. *Given four lines in general position; there is one and only one point from which the feet of the perpendiculars to the lines are concurrent, and this point is the common point of the four circumcircles.*

The line containing the feet of the perpendiculars from this common point is called the Simson line of the complete quadrilateral.

Exercise. *Prove that the centers of the four circumcircles also lie on a circle passing through their common point.*

198. Theorem.* *The area of the pedal triangle of a point P is proportional to the power of P with regard to the circumcircle:*

$$F = \tfrac{1}{2}(R^2 - \overline{OP}^2) \sin \alpha_1 \sin \alpha_2, \sin \alpha_3 = \frac{(R^2 - \overline{OP}^2)}{4\,R^2} \Delta$$

---

* The rest of the chapter may be omitted at the option of the reader.

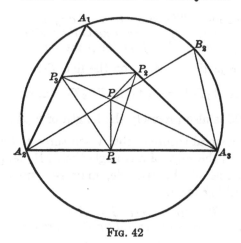

FIG. 42

Let A_2P meet the circumcircle at B_2; then

$$\angle A_2PA_3 = \angle P_2P_1P_3 + \angle A_2A_1A_3 = \angle A_2B_2A_3 + \angle B_2A_3P$$

whence

$$\angle P_2P_1P_3 = \angle B_2A_3P$$

Now

$$F = \text{area } P_2P_1P_3 = \tfrac{1}{2}\overline{P_1P_2}\cdot\overline{P_1P_3}\sin\angle P_2P_1P_3$$

$$= \tfrac{1}{2}\overline{P_1P_2}\cdot\overline{P_1P_3}\sin\angle B_2A_3P$$

$$= \tfrac{1}{2}\overline{PA_3}\sin\alpha_3\cdot\overline{PA_2}\sin\alpha_2\cdot\sin\angle B_2A_3P$$

But

$$\frac{\sin\angle B_2A_3P}{\sin\angle A_2B_2A_3} = \frac{\overline{PB_2}}{\overline{PA_3}}$$

whence

$$F = \tfrac{1}{2}\overline{PA_2}\cdot\overline{PB_2}\sin\angle A_2B_2A_3\cdot\sin\alpha_2\cdot\sin\alpha_3$$

$$= \tfrac{1}{2}(R^2 - \overline{OP^2})\sin\alpha_1\cdot\sin\alpha_2\cdot\sin\alpha_3$$

Corollaries:

a. The locus of a point whose pedal triangle has a given area is a circle concentric with the circumcircle; of points within the circle, the circumcenter has the greatest pedal triangle.

b. The area of the pedal triangle of a point on the circumcircle is zero.

c. The radius of the pedal circle of a point P is given by

$$r = \frac{\overline{A_1P} \cdot \overline{A_2P} \cdot \overline{A_3P}}{2(R^2 - \overline{OP}^2)}$$

For one of the formulas for the area of a triangle (**15** *d*) gives

$$\text{area } P_1P_2P_3 = \frac{\overline{P_2P_3} \cdot \overline{P_3P_1} \cdot \overline{P_1P_2}}{4r}$$

But $\overline{P_2P_3} = \overline{A_1P} \sin \alpha_1$, etc.; making these substitutions and replacing the area by the value given above, we have the result at once.

199. Theorem. *If the lines A_1P, A_2P, A_3P are extended to meet the circumcircle at B_1, B_2, B_3 respectively, then triangle $B_1B_2B_3$ is directly similar to the pedal triangle of P with respect to $A_1A_2A_3$.*

For $\quad \angle B_2B_1A_1 = \angle B_2A_2A_1 = \angle PP_1P_3$; etc.

Is the point P self-homologous in these similar triangles? (Cf. **244** *c*.)

200. Theorem. *If a given triangle is subjected to an inversion, the resulting triangle is directly similar to the pedal triangle of the center of inversion with regard to the given triangle.*

For if the triangle $B_1B_2B_3$ is inverse to $A_1A_2A_3$ with regard to a circle whose center is C, we have by **75**

$$\angle B_2B_1B_3 + \angle A_2A_1A_3 = \angle A_2CA_3$$

Again, if $C_1C_2C_3$ is the pedal triangle of C with regard to $A_1A_2A_3$,

$$\angle A_2CA_3 = \angle A_2A_1A_3 + \angle C_2C_1C_3$$

whence triangles $B_1B_2B_3$ and $C_1C_2C_3$ are directly similar. (Cf. also **133**.)

201. Theorem. *If two points P and P' are inverse with respect to the circumscribed circle of a triangle $A_1A_2A_3$, their pedal triangles are inversely similar.*

We know **75** that if O is the center of the circle,

$$\measuredangle A_2PA_3 + \measuredangle A_2P'A_3 = \measuredangle A_2OA_3 = 2\measuredangle A_2A_1A_3$$

But
$$\measuredangle A_2PA_3 = \measuredangle A_2A_1A_3 + \measuredangle P_2P_1P_3,$$
$$\measuredangle A_2P'A_3 = \measuredangle A_2A_1A_3 + \measuredangle P_2'P_1'P_3'$$

whence
$$\measuredangle P_2P_1P_3 + \measuredangle P_2'P_1'P_3' = 0$$

202. Theorem. *In this figure the distances from P and P' to the vertices of the triangle are proportional.*

For by similar triangles (cf. **95**) we prove at once that

$$\frac{\overline{OP}}{R} = \frac{R}{\overline{OP'}} = \frac{\overline{PA_1}}{\overline{P'A_1}} = \frac{\overline{PA_2}}{\overline{P'A_2}} = \frac{\overline{PA_3}}{\overline{P'A_3}}$$

203. Theorem. *The points in which the perpendicular bisector of a side of a triangle meets the other sides are inverse with regard to the circumcircle.*

204. Theorem. *If a set of four points is subjected to an inversion, the pedal triangle of one with regard to the triangle of the other three is inversely similar to the corresponding pedal triangle in the inverse figure.*

This remarkable result is derived easily from **75** and **186**.

205. Problem. *To determine a point P whose pedal triangle $P_1P_2P_3$ with regard to a given triangle $A_1A_2A_3$ shall be similar to a given triangle $C_1C_2C_3$.*

Assuming that $P_1P_2P_3$ is to be directly similar to $C_1C_2C_3$, the position of P is uniquely determined by the equations

$$\measuredangle A_2PA_3 = \measuredangle A_2A_1A_3 + \measuredangle C_2C_1C_3, \text{ etc.}$$

In practice, the simplest construction is to connect any points D_1 and D_2 on A_2A_3 and A_3A_1 respectively; then construct triangle $D_1D_2D_3$ similar to $C_1C_2C_3$. Let A_3D_3 meet

A_1A_2 at Q_3; and draw Q_3Q_1 and Q_3Q_2 parallel to D_3D_1 and D_3D_2. Since $Q_1Q_2Q_3$ is similar to $C_1C_2C_3$, its Miquel point will be the desired point P.

206. From another point of view, this problem is equivalent to that of finding a point whose distances from the vertices of the triangle are in given ratios (**95**). The distances from the vertices of the triangle to a point P are determined when the form of the pedal triangle is given. For

$$\overline{A_1P} = \frac{\overline{P_2P_3}}{\sin \alpha_1}, \text{ etc.}$$

Conversely, if the ratios $\overline{PA_1} : \overline{PA_2} : \overline{PA_3}$ are given, then the ratios $\overline{P_2P_3} : \overline{P_3P_1} : \overline{P_1P_2}$ can be determined; a Miquel triangle can be inscribed, and the point P can be found as in the foregoing. In this case there are two possible positions of P, according as $P_1P_2P_3$ is directly or inversely similar to $C_1C_2C_3$. This agrees with the result found in **95**, and we can now not only observe the solution of the problem there stated, but we can state exactly the conditions under which the problem can be solved.

Theorem. *There can be found two points P, P' whose distances from three given points A_1, A_2, A_3 are proportional to given lengths p_1, p_2, p_3, provided that the products $p_1 \cdot \overline{A_2A_3}$, $p_2 \cdot \overline{A_3A_1}$, $p_3 \cdot \overline{A_1A_2}$ can be made the sides of a triangle. These points are inverse with regard to the circumcircle of $A_1A_2A_3$.*

We conclude this chapter with a number of exercises and simple applications.

207. Theorem (Mannheim, *Educ. Times*, 1890). *In the Miquel figure (**184**), let any three concurrent lines A_1M, A_2M, A_3M, meet the respective Miquel circles at X_1, X_2, X_3, then X_1, X_2, X_3, M, P lie on a circle.*

For $\qquad \angle PX_1M = \angle PP_2A_1$, etc.

$$\angle PX_1M = \angle PX_2M = \angle PX_3M = \angle PP_1, a_1$$

208. The foregoing proof fails when M is at infinity. We may prove independently:

Theorem. *If parallel lines through the vertices meet the corresponding Miquel circles at Y_1, Y_2, Y_3, then these points are on a line through the Miquel point P.*

209. Theorem. *Conversely, if any circle through P cuts the Miquel circles at X_1, X_2, X_3, respectively, then A_1X_1, A_2X_2, A_3X_3 meet at a point M on the circle; and if any line through P cuts the circles at Y_1, Y_2, Y_3, then A_1Y_1, A_2Y_2, A_3Y_3 are parallel.*

210. Theorem. *If the perpendiculars to the sides of a triangle from a point on the circumcircle are extended to meet the circle again, the intersections form a triangle congruent to the given triangle.*

211. Theorem. *If three circles meet at a point concyclic with their centers, their other intersections are collinear.*

212. Theorem. *Given a line and a point P not on it. Through points A_1, A_2, A_3, ... of the line, draw lines A_1X_1, A_2X_2, A_3X_3, ... perpendicular respectively to PA_1, PA_2, PA_3, ... The circle circumscribed about any triangle whose sides are three of these lines A_1X_1, A_2X_2, A_3X_3, ... including the given line itself, passes through P.*

Exercise. Complete the proofs of all unproved propositions in this chapter, namely: **185, 187, 188** *a-e*, **190** *a-c*, **193, 194** *a-c*, **197, 198** *a, b*, **203, 204, 205, 207–212**.

CHAPTER VIII

THEOREMS OF CEVA AND MENELAUS

213. Many of the most interesting theorems of the triangle are concerned with sets of lines, one through each vertex of the triangle, which are concurrent. Obvious examples are the medians, the altitudes, and the angle bisectors. Other theorems deal with sets of points, one on each side of the triangle, which are collinear. In the present chapter we set up general criteria as to the concurrence of such triads of lines, or the collinearity of such triads of points. As immediate consequences of these theorems we shall notice some results already established; and a large number of further theorems will be deduced.

214. Theorem. *If three lines from the vertices of a triangle are concurrent at P, and meet the opposite sides at P_1, P_2, P_3 respectively, then*

$$\frac{\overline{P_1A_2} \cdot \overline{P_2A_3} \cdot \overline{P_3A_1}}{\overline{P_1A_3} \cdot \overline{P_2A_1} \cdot \overline{P_3A_2}} = -1$$

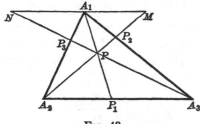

Fig. 43

Through A_1 draw MN parallel to A_2A_3; let it meet A_2P at M, A_3P at N. Then by similar triangles,

$$\frac{\overline{P_1A_2}}{\overline{P_1A_3}} = \frac{\overline{A_1M}}{\overline{A_1N}}, \quad \frac{\overline{P_2A_3}}{\overline{P_2A_1}} = \frac{\overline{A_3A_2}}{\overline{A_1M}}, \quad \frac{\overline{P_3A_1}}{\overline{P_3A_2}} = \frac{\overline{A_1N}}{\overline{A_2A_3}}$$

Multiplying,

$$\frac{\overline{P_1A_2}\cdot\overline{P_2A_3}\cdot\overline{P_3A_1}}{\overline{P_1A_3}\cdot\overline{P_2A_1}\cdot\overline{P_3A_2}} = \frac{\overline{A_2A_3}}{\overline{A_3A_2}} = -1$$

It is evident that the sign is necessarily negative, since for any position of P an odd number of the ratios will be negative. The proof is valid for every position of P, provided P_1, P_2, P_3 are actual points. If one or more of them be at infinity, we replace the corresponding ratio or ratios by $+1$, as explained in **10**, and the proof is easily adapted to this case. Again, P itself may be at infinity; that is, the theorem is valid for a set of parallels through the vertices of a triangle. This theorem furnishes an excellent illustration of the advantages of the convention as to points at infinity; without this convention, it would resolve into a number of cases, requiring separate proofs, and with embarrassing exceptions.

215. Theorem. *Conversely, if P_1, P_2, P_3 are so chosen on the sides of a triangle that*

$$\frac{\overline{P_1A_2}\cdot\overline{P_2A_3}\cdot\overline{P_3A_1}}{\overline{P_1A_3}\cdot\overline{P_2A_1}\cdot\overline{P_3A_2}} = -1$$

then the lines A_1P_1, A_2P_2, A_3P_3 are concurrent.

For in the first place if no two of these three lines meet, the three are concurrent at infinity, and the theorem is proved. Otherwise, let A_1P_1 and A_2P_2 meet at P; then let A_3P and A_1A_2 meet at Q_3. Then by **214**,

$$\frac{\overline{P_1A_2}\cdot\overline{P_2A_3}\cdot\overline{Q_3A_1}}{\overline{P_1A_3}\cdot\overline{P_2A_1}\cdot\overline{Q_3A_2}} = -1$$

whence

$$\frac{\overline{P_3A_1}}{\overline{P_3A_2}} = \frac{\overline{Q_3A_1}}{\overline{Q_3A_2}}$$

and therefore P_3 coincides with Q_3. Thus A_1P_1, A_2P_2, A_3P_3 are concurrent at P.

216. Combining these two results, we have the famous

Theorem of Ceva. *A necessary and sufficient condition that lines from the vertices of a triangle $A_1A_2A_3$ to points P_1, P_2, P_3 on the opposite sides be concurrent is that*

$$\frac{\overline{P_1A_2} \cdot \overline{P_1A_3} \cdot \overline{P_3A_1}}{\overline{P_1A_3} \cdot \overline{P_2A_1} \cdot \overline{P_3A_2}} = -1$$

or, what is the same thing (**85**):

$$\frac{\sin \angle P_1A_1A_2 \cdot \sin \angle P_2A_2A_3 \cdot \sin \angle P_3A_3A_1}{\sin \angle P_1A_1A_3 \cdot \sin \angle P_2A_2A_1 \cdot \sin \angle P_3A_3A_2} = -1$$

217. A familiar version of this theorem is:

Theorem. *Three concurrent lines from the vertices of a triangle divide the opposite sides in such fashion that the product of three non-adjacent segments equals the product of the other three.*

218. Related to the foregoing theorem is that of Menelaus:

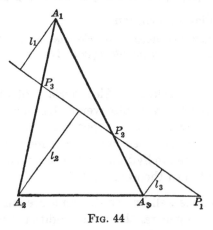

FIG. 44

Theorem. *Three points P_1, P_2, P_3 on the sides of a triangle $A_1A_2A_3$ are collinear if and only if*

$$\frac{\overline{P_1A_2} \cdot \overline{P_2A_3} \cdot \overline{P_3A_1}}{\overline{P_1A_3} \cdot \overline{P_2A_1} \cdot \overline{P_3A_2}} = 1$$

Suppose first that the points are collinear. Let perpendiculars be dropped to the line from A_1, A_2, A_3, and denote their lengths by l_1, l_2, l_3 respectively. Then, signs disregarded, we have

$$\frac{\overline{P_1A_2}}{\overline{P_1A_3}} = \frac{l_2}{l_3}, \qquad \frac{\overline{P_2A_3}}{\overline{P_2A_1}} = \frac{l_3}{l_1}, \qquad \frac{\overline{P_3A_1}}{\overline{P_3A_2}} = \frac{l_1}{l_2}$$

so that the product of the ratios is either $+1$ or -1. But a line must cut internally either two sides of a triangle, or none; therefore in every case the product of the ratios is positive. There are a few special cases, which present no difficulties; and the converse is proved by the indirect method used in **215**.

Corollary. As before, we have the alternative form

$$\frac{\sin \angle P_1A_1A_2 \cdot \sin \angle P_2A_2A_3 \cdot \sin \angle P_3A_3A_1}{\sin \angle P_1A_1A_3 \cdot \sin \angle P_2A_2A_1 \cdot \sin \angle P_3A_3A_2} = 1$$

and the familiar enunciation:

Any line cuts the sides of a triangle so that the product of three non-adjacent segments equals the product of the other three.

Historical. Menelaus of Alexandria (not to be confused with him of Sparta), who discovered this theorem, flourished about 100 B.C., and wrote on geometry and trigonometry. His theorem was forgotten until it was rediscovered by Giovanni Ceva, an Italian hydraulic engineer and mathematician, who published both theorems in 1678.

219. Numerous well-known theorems are immediate consequences of these general theorems; additional theorems may be obtained with equal ease.

a. The medians of a triangle are concurrent.
b. The altitudes are concurrent.
c. The bisectors of the interior angles are concurrent.

d. The bisector of any interior angle and those of the other two exterior angles, are concurrent.

e. The exterior angle bisectors meet the opposite sides in collinear points.

f. The bisectors of two interior angles and the third exterior angle, meet the sides in collinear points.

g. If P_1 is a point "halfway around the triangle" from A_1, so that

$$\overline{A_1A_2} + \overline{A_2P_1} = \overline{P_1A_3} + \overline{A_3A_1}$$

and if P_2 and P_3 are similarly located, then A_1P_1, A_2P_2, A_3P_3 are concurrent. (Nagel point, **291** *b*, **361**.)

For we easily find that $\overline{P_1A_2} = s - a_3$, $\overline{P_1A_3} = a_2 - s$, etc.

220. Theorem. *If three concurrent lines A_1P, A_2P, A_3P meet the opposite sides of a triangle at P_1, P_2, P_3; and if P_2P_3 meets A_2A_3 at Q_1, then P_1 and Q_1 divide A_2A_3 internally and externally in the same ratio, or P_1, Q_1, A_2, A_3 are a harmonic set of points* (**87**).

For

$$\frac{\overline{P_1A_2} \cdot \overline{P_2A_3} \cdot \overline{P_3A_1}}{\overline{P_1A_3} \cdot \overline{P_2A_1} \cdot \overline{P_3A_2}} = -1,$$

while

$$\frac{\overline{Q_1A_2} \cdot \overline{P_2A_3} \cdot \overline{P_3A_1}}{\overline{Q_1A_3} \; \overline{P_2A_1} \; \overline{P_3A_2}} = 1$$

hence

$$\frac{\overline{P_1A_2}}{\overline{P_1A_3}} = -\frac{\overline{Q_1A_2}}{\overline{Q_1A_3}}$$

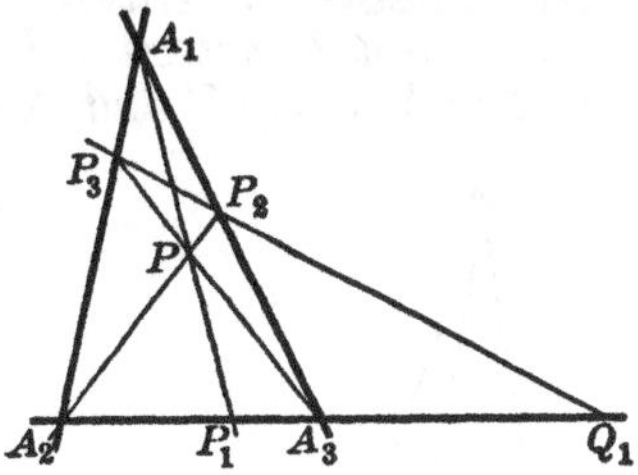

Fig. 45

Other versions of the same relation:

221. Theorem. *In a complete quadrangle $A_2A_3P_2P_3$, any connector, as A_2A_3, is divided harmonically by the diagonal point lying on it, namely Q_1, and the point P_1 where it is cut by the line through the other diagonal points A_1 and P.*

222. Theorem. *In a complete quadrilateral whose sides are A_1A_2, A_2A_3, A_3A_1, P_3Q_1, the line connecting a vertex A_1 with a point of intersection P of two diagonals cuts an opposite side $A_2A_3Q_1$ harmonically (at P_1).*

These two theorems are important in projective geometry.

223. Theorem.   *If A_1P_1, A_2P_2, A_3P_3 are concurrent at P, and if P_2P_3, P_3P_1, P_1P_2 meet A_2A_3, A_3A_1, A_1A_2 at Q_1, Q_2, Q_3 respectively, then Q_1, Q_2, Q_3 are collinear.*

This follows at once from **220**.   The line $Q_1Q_2Q_3$ is called the *trilinear polar* of P.

224. Theorem.   *With the same hypotheses the lines A_1P_1, A_2Q_2, A_3Q_3 are concurrent at a point P'.*

Thus with any point P of the plane, not on any side of the triangle, there are associated three points P', P'', P'''; the four points usually have interesting common properties.   A familiar example consists of the points of intersection of the bisectors of the angles of a triangle.

Corollary.   *The perpendiculars to the sides of the triangle from the four associated points P, P', P'', P''' are proportional in numerical value, but differ as to signs.*

225. Theorem.   *Given two fixed lines AM and AN and a fixed point B not on either; let any two lines through B cut AM at M and M' and AN at N and N' respectively.   Then MN' and M'N meet at a point X whose locus is a straight line through A.   Moreover the line BX is cut harmonically by AM and AN.*

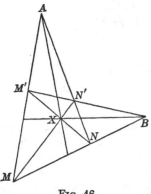

FIG. 46

This follows at once from **220**, identifying the triangle AMN with $A_1A_2A_3$ and B with Q_1; it is easily proved that every line through B is cut harmonically by the three lines from A.

Thus every point not on either of two lines has a *line polar* with regard to the two lines (cf. **138, 139**).

226. Theorem.   *If a line B_2B_3 is parallel to A_2A_3, and A_2B_2 meets A_3B_3 at P, then A_1P bisects A_2A_3 and B_2B_3.*

This may be regarded as a special case of the foregoing theorem; it is proved immediately by Menelaus's theorem.

Theorem. *The line joining the mid-points of the parallel sides of a trapezoid passes through the point of intersection of the diagonals, and also that of the non-parallel sides.*

227. Theorem. *If a circle cuts the sides of a triangle at X_1, Y_1, X_2, Y_2, X_3, Y_3, and if A_1X_1, A_2X_2, A_3X_3 are concurrent, then A_1Y_1, A_2Y_2, A_3Y_3 are concurrent.*

For $\qquad \overline{X_1A_2} \cdot \overline{Y_1A_2} = \overline{X_3A_2} \cdot \overline{Y_3A_2}$, etc.

228. Theorem. *If two triangles $A_1A_2A_3$ and $B_1B_2B_3$ are inscribed in the same circle, and if A_1B_1, A_2B_2, A_3B_3 are concurrent, then, disregarding signs,*

$$\frac{\overline{A_1B_2} \cdot \overline{A_2B_3} \cdot \overline{A_3B_1}}{\overline{A_1B_3} \cdot \overline{A_2B_1} \cdot \overline{A_3B_2}} = 1$$

On account of possible ambiguity of sign, the converse of this theorem is not true; and the theorem may not be used, as some geometers have used it, to prove the concurrence of lines.

CENTERS OF SIMILITUDE OF THREE CIRCLES

229. We recall that the homothetic centers of similitude of two circles divide the line of centers internally and externally in the ratio of the radii. If then we have three circles, whose centers form a triangle, the following properties of the centers of similitude are derived at once from the theorems of Ceva and Menelaus (cf. also the analysis of **171**). Among these are the rather simple theorems which excited the wonder and admiration of Herbert Spencer.*

a. The external centers of similitude of three circles are collinear.

b. Any two internal centers of similitude are collinear with the third external one.

---

* Cf. *American Math. Monthly*, xxviii, May, 1921, p. 229.

c. If the center of each circle is connected with the internal center of similitude of the other three, the connectors are concurrent.

d. If one center is connected with the internal center of the other two, the others with the corresponding external centers, the connectors are concurrent.

230. Theorem. *The mid-points of the diagonals of a complete quadrilateral are collinear.*

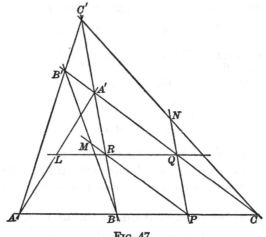

Fig. 47

This theorem has already been proved; but, as may well be supposed, a simple proof can be derived from the theorem of Menelaus. The following form of the proof is due to Hillyer.*

With the usual notation, let the middle points of BC, CA', $A'B$ be P, Q, R respectively. If QR cuts AA' at L, RP cuts BB' at M, and PQ cuts CC' at N, our task is to prove L, M, N collinear, since these are clearly the middle points of the respective diagonals. We have the following proportions:

$$\frac{\overline{LQ}}{\overline{LR}} = \frac{\overline{AC}}{\overline{AB}}, \qquad \frac{\overline{MR}}{\overline{MP}} = \frac{\overline{B'A'}}{\overline{B'C}}, \qquad \frac{\overline{NP}}{\overline{NQ}} = \frac{\overline{C'B}}{\overline{C'A'}}$$

* Durell, *Modern Geometry*, p. 85.

But since $AB'C'$ is a transversal cutting the sides of triangle $A'BC$, the product of the right-hand members is $+1$; and if the product of the left-hand members is $+1$, then the points L, M, N on the sides of triangle PQR are collinear.

ISOGONAL CONJUGATES

231. We now introduce a relationship by means of which the points of the plane of a triangle are associated in pairs. To every point there corresponds a conjugate point; and among the pairs of partners we shall later find some of the points in which we are most interested.

Definition. If two rays through the vertex of an angle make equal angles with its sides, they are said to be *isogonal*. In other words, two rays are isogonal with regard to an angle if they are symmetrical with regard to the bisector of that angle.

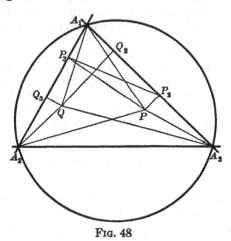

Fig. 48

Each line through the vertex of an angle has a definite isogonal; each angle bisector is self-isogonal. The fundamental theorem of isogonals is the following:

232. Theorem. *If three lines from the vertices of a triangle are concurrent, their isogonals are also concurrent.*

The proof is an immediate application of the theorem of Ceva. If A_1P_1 and A_1Q_1 are isogonal in triangle $A_1A_2A_3$, then

$$\frac{\sin \angle A_2A_1P_1}{\sin \angle A_3A_1P_1} = \frac{\sin \angle Q_1A_1A_3}{\sin \angle Q_1A_1A_2}, \text{ etc.}$$

Definition. Two points P, Q in the plane of a triangle $A_1A_2A_3$, such that

$$\angle A_2A_1P = \angle QA_1A_3,$$
$$\angle A_3A_2P = \angle QA_2A_1,$$
$$\angle A_1A_3P = \angle QA_3A_2$$

are *isogonally conjugate* with regard to the triangle.

233. Exercise. By virtue of the foregoing theorem every point in the plane, in general, has a definite isogonal conjugate. Choose points at random in the plane, and by freehand sketching find the position of the isogonal conjugate of each. For example, trace the position of a point, as its isogonal conjugate traces a straight line; a circle; especially a circle through two vertices of the triangle.

Show that the isogonal conjugate of any point on a side of the triangle is at the opposite vertex; and as a moving point approaches a vertex from any direction, its isogonal conjugate approaches a limiting position on the opposite side.

234. Theorem. *The isogonal conjugate of a point on the circumcircle is at infinity; and conversely.*

For if P is on the circumcircle, and the isogonals of A_1P and A_2P meet the circle at P_1' and P_2' respectively. Then by definition, arcs A_3P and $P_1'A_2$ are equal, and so on. Hence we easily prove by equal arcs that A_1P_1' and A_2P_2' are parallel. Conversely, we easily prove that the isogonals of a set of parallels through the vertices are concurrent at a point on the circumcircle.

In general, then, the points of the plane are paired off into isogonal conjugates; but the partner of each point of the circumcircle is at infinity, and each vertex is polygamous with all the points of the opposite side.   Every point not on the circle nor on any side of the given triangle itself has an actual conjugate; in particular, each of the four points of intersection of the angle bisectors is its own conjugate; and these are the only self-conjugate points.

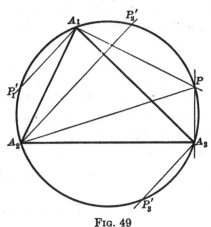

FIG. 49

235. Theorem. *The distances to the sides of an angle, from points on isogonal rays, are inversely proportional.*

In figure 48, we prove at once by similar triangles that $\overline{PP_2}\cdot\overline{PP_3} = \overline{QQ_2}\cdot\overline{QQ_3}$. The corollary is of some importance:

Corollary. *The perpendiculars to the sides of a triangle from isogonal conjugate points are inversely proportional,*

$$p_1q_1 = p_2q_2 = p_3q_3.$$

236. Theorem. *The feet of the perpendiculars from two isogonally conjugate points lie on a circle; that is, isogonal conjugates have a common pedal circle, whose center is midway between them.*

For if the pedal triangles of two isogonal conjugates P and Q (figure 48) are $P_1P_2P_3$ and $Q_1Q_2Q_3$, we have

$$\frac{\overline{A_2P_1}}{\overline{A_2P_3}} = \frac{\cos \angle PA_2P_1}{\cos \angle PA_2P_3} = \frac{\cos \angle Q_3A_2Q}{\cos \angle Q_1A_2Q} = \frac{\overline{A_2Q_3}}{\overline{A_2Q_1}}$$

whence

$$\overline{A_2P_1}\cdot\overline{A_2Q_1} = \overline{A_2P_3}\cdot\overline{A_2Q_3}$$

and each two pairs of points are concyclic. It follows by **62 a**
that the six points lie on one and the same circle.   The center
of the common pedal circle is evidently midway between the
two points.   For a point on the circumcircle we see that there
is, properly speaking, no pedal circle (**191**) and no isogonal
conjugate.

237. Theorem. *The sides of the pedal triangle of a point
are perpendicular to the connectors of the corresponding vertices with the isogonal conjugate.*

That is, if P and Q are isogonal conjugates, A_1Q is perpendicular to P_2P_3. For (figure 48) let these lines meet at R,
then since A_1, P_2, P_3, P lie on a circle,

$$\measuredangle RP_3A_1 = \measuredangle P_2PA_1$$

Also

$$\measuredangle PA_1P_2 = \measuredangle P_3A_1R$$

so that the triangles A_1P_2P and A_1RP_3 are directly similar,
and R is a right angle.

238. We have a fundamental angle formula for isogonal
conjugates, similar to the formulas for inversion (**75**) and for
Miquel points (**186**).

Theorem.   *If P and Q are isogonal conjugates,*

$$\measuredangle A_2PA_3 + \measuredangle A_2QA_3 = \measuredangle A_2A_1A_3$$

(If both points are within the triangle, this is equivalent to

$$\angle A_2PA_3 + \angle A_2QA_3 = 180° + \alpha_1.)$$

For
$$\measuredangle A_2P, PA_3 = \measuredangle A_2P, a_1 + \measuredangle a_1, PA_3$$
$$\measuredangle A_2Q, QA_3 = \measuredangle A_2Q, a_1 + \measuredangle a_1, QA_3$$
$$= \measuredangle a_3, A_2P + \measuredangle PA_3, a_2$$

(from the definition of isogonals)

Adding, $\measuredangle A_2PA_3 + \measuredangle A_2QA_3 = \measuredangle a_3, a_1 + \measuredangle a_1, a_2 = \measuredangle A_2A_1A_3$

Corollary. *If a point traces a circle through two vertices of the triangle, so that* $\angle A_2PA_3$ *is constant, then its isogonal conjugate traces another circle through the same two points.*

239. Theorem. *If any circle cuts the sides of a triangle at* $P_1, Q_1, P_2, Q_2, P_3, Q_3$ *in any order, then*

$$\angle P_2P_1P_3 + \angle Q_2Q_1Q_3 + \angle A_2A_1A_3 = 0$$

For $\angle P_2P_1P_3 = \angle P_2Q_2P_3 = \angle A_3A_1, A_1A_2 + \angle A_1A_2, P_3Q_2$
$$= \angle A_3A_1A_2 + \angle Q_3Q_1Q_2$$

Corollary. *In particular, these equations characterize the pedal triangles of any two isogonal conjugate points P and Q.*

240. Theorem. *If any circle cuts the sides of a triangle, at* $P_1, Q_1, P_2, Q_2, P_3, Q_3$*, the Miquel points P and Q of the triads* $P_1P_2P_3$ *and* $Q_1Q_2Q_3$ *are isogonal conjugates.*[*]

For we have

$$\angle A_2QA_3 = \angle A_2A_1A_3 + \angle Q_2Q_1Q_3$$
$$\angle A_2PA_3 = \angle A_2A_1A_3 + \angle P_2P_1P_3$$
$$\angle A_2PA_3 + \angle A_2QA_3 = \angle A_2A_1A_3$$
$$+ [\angle P_2P_1P_3 + \angle Q_2Q_1Q_3 + \angle A_2A_1A_3]$$

But we have just seen that the sum inside the bracket is zero, hence we have the conditions that P and Q be isogonal conjugates,

$$\angle A_2PA_3 + \angle A_2QA_3 = \angle A_2A_1A_3, \text{ etc.}$$

ISOTOMIC CONJUGATES AND OTHER RELATIONS

241. Another relation very similar to that of isogonal conjugates, but far less important, is defined by the following theorem.

Theorem. *If three lines from the vertices of a triangle, and concurrent at* P*, meet the opposite sides at* P_1, P_2, P_3 *re-*

---

[*] Lachlan, *l.c.*, p. 133, 9, 10; Gallatly, *l.c.*, p. 110; Barrow, *American Math. Monthly*, 1913, p. 251.

spectively; and if we cut off A_2Q_1, A_3Q_2, A_1Q_3 equal respectively to P_1A_3, P_2A_1, P_3A_2, then A_1Q_1, A_2Q_2, A_3Q_3 are concurrent at a point Q, called the isotomic conjugate of P.

The proof, based on the theorem of Ceva, is obvious.

242. Theorem.   *There are four points, each of which is isotomically self-conjugate; namely the median point M, and each of the points of intersection of lines through the vertices parallel to the opposite sides.*   (Cf. **277**.)

243. Theorem.   *If a line cuts the sides of a triangle at P_1, P_2, P_3, and if the isogonals of A_1P_1, A_2P_2, A_3P_3 are A_1Q_1, A_2Q_2, A_3Q_3, then Q_1, Q_2, Q_3 are collinear; and if R_1, R_2, R_3 are isotomic to P_1, P_2, P_3 on the respective sides, they are also collinear.*

244. A few additional theorems and exercises in isogonals and isotomics are suggested.

a. The product of the ratios in which two isogonals from a given vertex of a triangle cut the opposite side is constant, and equal to the ratio of the squares of the adjacent sides (**84**),

$$\frac{\overline{P_1A_2}}{\overline{P_1A_3}} \cdot \frac{\overline{Q_1A_2}}{\overline{Q_1A_3}} = \left(\frac{\overline{A_1A_2}}{\overline{A_1A_3}}\right)^2$$

b. If a given point is reflected with regard to the sides of a given triangle, the center of the circle through the reflections is the isogonal conjugate of the given point (**236**).

c. We saw in **199** that if two triangles are inscribed in a circle and lines connecting their vertices meet at a point P, either is similar to the pedal triangle of P in the other. Show further that *P, referred to one of the similar triangles, is homologous to its isogonal conjugate in the other.*

d. An altitude of a triangle and a radius of the circumcircle from the same vertex are isogonal.

e. The product of the lengths of two isogonals from a vertex of a triangle, the one measured to the opposite side and the other to the circumcircle, is equal to the product of the including sides.

As special cases of this theorem we may notice the theorems of **101** and (with a little manipulation) **99**.

f. Construct a triangle, if one side is on a given line, the other sides pass through given points, and two other given points are isogonal conjugates.

MISCELLANEOUS EXERCISES

245. Theorem. *Let A_1P_1, A_2P_2, A_3P_3 be concurrent at P. Let P_1Q_2 be drawn parallel to a_3, P_2Q_3 parallel to a_1, and P_3Q_1 to a_2. Then A_1Q_1, A_2Q_2, A_3Q_3 are concurrent.*
Similarly if P_1R_3 is parallel to a_2, etc., then A_1R_1, A_2R_2, A_3R_3 are concurrent, at a point R.
In the same figure, if M is the point of intersection of the medians, A_1P, A_2M, A_3Q are concurrent; A_1P, A_2R, A_3M are concurrent; and so on.
The triangles $P_1P_2P_3$, $Q_1Q_2Q_3$, $R_1R_2R_3$ have equal areas. (Cf. **107.***)*

246. By applying the theorems of Ceva and Menelaus to the triangle $O_1O_2O_3$, whose vertices are the mid-points of the sides, we may obtain a multiplicity of theorems, of which the following are typical:

If three concurrent lines are drawn from the vertices of the given triangle to meet the opposite sides, and the mid-points of these connectors are joined to the mid-points of the sides, the lines so drawn are also concurrent.
If P_1, P_2, P_3 are collinear points on the sides of a triangle, the mid-points of the lines A_1P_1, A_2P_2, A_3P_3 are also collinear.
Through the mid-points of the sides, let lines be drawn parallel to three given concurrent lines through the vertices. These are concurrent, and the two points of concurrence are homologous in the similar triangles $A_1A_2A_3$, $O_1O_2O_3$, so that the line joining them is trisected by the median point M.

247. We may deal in the same way with any triangle $P_1P_2P_3$ inscribed in the given triangle. For instance:

If A_1P_1, A_2P_2, A_3P_3 are concurrent at P, and if X_1, X_2, X_3 are the mid-points of P_2P_3, P_3P_1, P_1P_2 respectively, then

A_1X_1, A_2X_2, A_3X_3 *are concurrent.* (Use **84**.) *Also* O_1X_1, O_2X_2, O_3X_3 *are concurrent.*

More generally, if Y_1, Y_2, Y_3 *are any points on* P_2P_3, P_3P_1, P_1P_2, *such that* P_1Y_1, P_2Y_2, P_3Y_3 *are concurrent, then* A_1Y_1, A_2Y_2, A_3Y_3 *are concurrent.*

Exercise.   Give the complete proofs of the unproved propositions in this chapter, in the following sections: **219**, **221-228**, **229**, **235**, **239**, **241-244**, **245-247**.

CHAPTER IX

THREE NOTABLE POINTS

248. This chapter deals with the properties of three of the notable points associated with the triangle, whose history extends back to the ancient Greeks. These points, which are somewhat intimately related to one another, are the circumcenter O, the intersection of the perpendicular bisectors of the sides of the triangle, and center of the circumscribed circle; the orthocenter H, the intersection of the altitudes; and the median point M, the intersection of the medians. The notation has already been explained (**13**). It is true that the inscribed circle and its center were also known to the ancients; but their properties are best discussed separately and are the subject of Chapter X.

249. We observe first that the median point M is always within the triangle, and trisects each median. If all angles of the triangle are acute, the orthocenter and circumcenter are also within the triangle; but if angle A_1 is obtuse, the orthocenter lies outside the triangle on the extension of altitude H_1A_1, while the circumcenter is on the extension of OO_1 beyond A_2A_3. In a right triangle the orthocenter is at the vertex of the right angle, while the circumcenter is the midpoint of the hypotenuse.

250. Theorem. *The angles subtended at the circumcenter by the sides are double the angles of the triangle:*

$$\angle A_2OA_3 = 2\,\alpha_1, \quad \angle A_2OO_1 = \alpha_1;$$

except that if A_1 is obtuse,

$$\angle A_2OA_3 = 2\,(180° - \alpha_1), \quad A_2OO_1 = 180° - \alpha_1$$

In any case,

$$\angle A_2 O A_3 = 2 \angle A_2 A_1 A_3, \quad \angle A_2 O O_1 = \angle A_2 A_1 A_3$$

These formulas may be derived directly from the figure, or deduced from **186**. The equation last given is the most useful form.

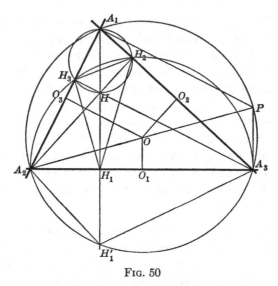

FIG. 50

251. Theorem. *The points H_2 and H_3 lie on the circle drawn on $A_2 A_3$ as diameter, and also on the circle on $A_1 H$ as diameter.*

252. From the foregoing theorems we can derive a number of formulas, expressing the values of the various angles and lines in the figure.

a. $\angle A_1 A_2 H = \angle A_1 A_3 H = 90° - \alpha_1;$ that is,
         $\angle A_1 A_2 H + \angle A_3 A_1 A_2 = 90°$

b. $\angle A_2 H H_3 = \angle A_3 A_1 A_2$

c. $\angle A_1 H_2 H_3 = \angle A_1 H H_3 = \angle A_3 A_2 A_1$

d. $\qquad \angle HH_1H_3 = \angle HA_2A_1 = \angle A_1A_3H = \angle H_2H_1H$

e. $\qquad \overline{A_2H_1} = a_3 \cos \alpha_2; \quad \overline{A_1H} = 2R \cos \alpha_1$

f. $\qquad \overline{OO_1} = R \cos \alpha_1; \quad \overline{A_1H}^2 + \overline{A_2A_3}^2 = 4R^2$

g. $\qquad \overline{HH_1} = 2R \cos \alpha_2 \cos \alpha_3;$

$\qquad h_1 = \overline{A_1H_1} = a_2 \sin \alpha_3 = a_3 \sin \alpha_2$

253. Translating into words some of these relations, we have the following propositions:

a. **Theorem.** *The circumcenter O and the orthocenter H are isogonal conjugates.*

b. **Theorem.** *Triangles $A_1A_2A_3$ and $A_1H_2H_3$ are inversely similar.*

c. **Theorem.** *The circumcenter O is the orthocenter of its own pedal triangle.*

d. **Theorem.** *The altitudes and sides of the given triangle bisect the interior and exterior angles of the triangle $H_1H_2H_3$.*

e. **Theorem.** *The radius A_1O is perpendicular to H_2H_3.*

f. **Theorem.** *The circles on A_2A_3 and A_1H as diameters are orthogonal to each other at H_2 and H_3.*

For if H_3T is tangent to the circle on A_1H as diameter,

$$\angle A_2H_3T = \angle A_1H_2H_3 = \angle A_3A_2H_3$$

so that if T is on A_2A_3, H_3A_2T is an isosceles triangle. It follows that T is at O_1. (Cf. **62** *f.*)

254. Theorem. *The segment of an altitude from the orthocenter to the side equals its extension from the side to the circumcircle; if A_1H is extended to meet the circumcircle at H_1' then $\overline{H_1H_1'} = \overline{HH_1}$.*

For at once triangles H_1HA_2 and $H_1H_1'A_2$ are proved congruent. In other words, the reflections of H with regard to the sides lie on the circumcircle.

255. Theorem. *The products of the segments of the respective altitudes are equal,*

$$\overline{HA_1} \cdot \overline{HH_1} = \overline{HA_2} \cdot \overline{HH_2} = \overline{HA_3} \cdot \overline{HH_3}$$

Several proofs are obvious. The constant product represents the power of H with regard to each of the circles on the sides as diameters; or half the power of H with regard to the circumcircle, in the light of **254**; or trigonometrically we find

$$\overline{HA_1} \cdot \overline{HH_1} = 2\,R\cos\alpha_1 \cdot 2\,R\cos\alpha_2 \cdot \cos\alpha_3 = 4\,R^2\cos\alpha_1\cos\alpha_2\cos\alpha_3$$

Corollary. $\overline{HA_1} \cdot \overline{HH_1} = \frac{1}{2}(a_1^2 + a_2^2 + a_3^2) - 4\,R^2$

For $\overline{HA_1} \cdot \overline{HH_1} = \overline{O_1A_2}^2 - \overline{O_1H}^2$

being the power of H with regard to the circle $O_1\,(O_1A_2)$.

But $\overline{O_1H}^2 + \frac{1}{4}\overline{A_2A_3}^2 = \frac{1}{2}(\overline{HA_2}^2 + \overline{HA_3}^2)$ **(96)**

and $\overline{HA_2}^2 = 4\,R^2 - \overline{A_1A_3}^2$, etc.

256. Theorem. *A chord of the circumcircle, perpendicular to one side of the triangle at an extremity, is equal to the segment from the orthocenter to the vertex opposite this side.*

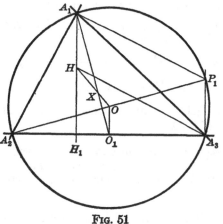

For if A_3P_1, perpendicular to A_2A_3, meets the circumcircle at P_1, then A_2P_1 is a diameter of the circle, and $A_2A_1P_1$ is also a right angle; P_1A_1 is parallel to HA_3, and $P_1A_1HA_3$ is a parallelogram. Hence its opposite sides P_1A_3 and A_1H are equal.

Fig. 51

Corollary. $\overline{A_1H} = 2\,\overline{OO_1}$.

For $\overline{A_3P_1} = 2\,\overline{O_1O}$; this may also be seen directly from **252** e, f.

257. Theorem (Euler). *The circumcenter, orthocenter, and median point of a triangle are collinear, and the last trisects the line joining the other two:* $2\overline{OM} = \overline{MH}$.

For if A_1O_1 and OH meet at X, triangles A_1HX and O_1OX are similar, being mutually equiangular. But

$$\overline{A_1H} = 2\overline{OO_1}$$

therefore $\qquad \overline{A_1X} = 2\overline{XO_1}, \ \overline{HX} = 2\overline{XO}$

showing that X trisects the median and is the median point. This well-known theorem has already been foreshadowed; we see, indeed, that M is the center of similitude of the directly similar triangles $A_1A_2A_3$ and $O_1O_2O_3$, whose orthocenters are respectively O and H. Other properties of the Euler line OMH will be indicated from time to time.

258. The Nine Point Circle. Since O and H are isogonal conjugates, it follows (**236**) that they have a common pedal circle; in other words, the feet of the altitudes and the mid-points of the sides lie on a circle. The center of this circle is midway between O and H; its radius is half that of the circumcircle, and it passes also through the mid-points of A_1H, A_2H, A_3H. It is called the nine point circle, and has such striking properties as to deserve a separate chapter. We therefore postpone until Chapter XI further investigation of this circle, which would otherwise be appropriate at this time.

ORTHOCENTRIC SYSTEMS

259. Definition. An *orthocentric system* is a set of four points, one of which is the orthocenter of the triangle of the other three.

Theorem. *In an orthocentric system, each point is the orthocenter of the triangle of the other three.*

For if H is the orthocenter of triangle $A_1A_2A_3$, the altitudes of triangle A_2A_3H are precisely A_2H_3, A_3H_2, A_1H_1, and these obviously are concurrent at A_1.

By this theorem, the four points are endowed with equal

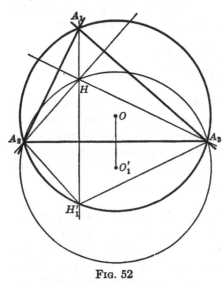

rank. Any three points not in a line therefore determine an orthocentric system; and the four points will be distinct except when three of them are vertices of a right triangle. In that case the fourth coincides with the vertex of the right angle; in all other cases, one point falls within the triangle of the other three, one of the four triangles is acute and the other three are obtuse.

FIG. 52

260. Theorem.   *The four circumcircles of an orthocentric system are equal.*

For we have proved that HH_1 and H_1H_1' are equal; therefore triangles A_2A_3H and $A_2A_3H_1'$ are congruent, and their circumscribed circles are equal.   In other words:

Theorem.   *The circle through two vertices and the orthocenter is equal to the circumcircle.*

The converse theorem was discussed in **104**, where it was established that the intersections of four equal circles constitute an orthocentric system.

261. Theorem.   *The centers of the circumcircles of an orthocentric system form another orthocentric system congruent to the first.* (Cf. **104** a.)

For the centers O_1', O_2', O_3' of circles A_2A_3H, etc., are the reflections of O with regard to the sides. Hence $\overline{OO_1'}$ is twice $\overline{OO_1}$, and therefore equal and parallel to A_1H; and so for all the connectors. Thus each of the connectors of the four points O, O_1', O_2', O_3' is equal and parallel to the corresponding connector of H, A_1, A_2, A_3.

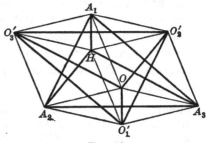

FIG. 53

Corollary. *The connectors of corresponding members of the two figures are concurrent, and bisect one another, at the mid-point F of OH. The twelve other connectors, as A_1O_2', HO_1', etc., are all equal to R, and are parallel to the three fixed directions A_1O, A_2O, A_3O.*

This figure may be regarded as the picture, or plane projection, of a solid figure, namely a parallelepiped held in such a position that the projections of its edges are equal. Then homologous points of the two orthocentric systems represent diagonally opposite points of the solid, and the twelve connectors the edges. The special case of an equilateral triangle and its center corresponds to a cube projected on a plane perpendicular to one diagonal.

262. Theorem. *The sum of the squares of any non-adjacent pair of connectors of an orthocentric system equals the square of the diameter of the circumcircle (*252 f*).*

263. Another characteristic property of orthocentric systems is suggested by the fact already noted that A_1H and A_2A_3 are diameters of two orthogonal circles. We propose the

Problem. *To construct a triangle, given the base A_2A_3 and the feet of the corresponding altitudes H_2 and H_3.*

It is obviously necessary and sufficient that the points H_2 and H_3 lie on the circle whose diameter is A_2A_3. If this condition is satisfied, then A_2H_3 and A_3H_2 will intersect at a definite point A_1, and A_2H_2 and A_3H_3 at a point H. Then H is evidently the orthocenter of $A_1A_2A_3$, and we have an orthocentric system. Moreover, the circle on A_1H as diameter is orthogonal to the first circle. Thus we have the converse theorem (cf. **62** f):

Theorem. *If two circles intersect orthogonally at P and Q, and AB is a diameter of the one, let AP and BQ meet at C, AQ and BP at D. Then CD is a diameter of the second circle perpendicular to AB, and A, B, C, D are an orthocentric system. In other words, the extremities of mutually perpendicular diameters of two orthogonal circles form an orthocentric system.*

264. An interesting property of the orthocenter is that of all triangles inscribed in a given acute triangle, the one having the minimum perimeter is the pedal triangle $H_1H_2H_3$ of the orthocenter. For this theorem we need the following lemmas:

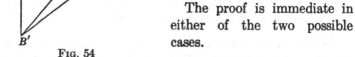

Theorem. *If the sides of a triangle are three non-concurrent angle bisectors of a second triangle, the vertices of the latter are feet of the altitudes of the first.*

The proof is immediate in either of the two possible cases.

Fig. 54

Theorem. *The shortest path joining two given points on one side of a line, and meeting this line, is a broken line whose parts make equal angles with the given line.*

The proof is obvious from the figure; this is a familiar exercise in elementary geometry.

Theorem. *Of all triangles inscribed in a given acute triangle, the triangle $H_1H_2H_3$ has the minimum perimeter.*

For * if $P_1P_2P_3$ is inscribed in $A_1A_2A_3$, and P_1P_2 and P_1P_3 do not make equal angles with A_2A_3, then if Q_1 is so located that P_2Q_1 and P_3Q_1 do make equal angles with A_2A_3, we have perimeter $P_2P_3Q_1$ less than that $P_2P_3P_1$. Thus if a triangle of minimum perimeter exists, its sides make equal angles with the sides of the given triangle, and must, as shown above, be the triangle $H_1H_2H_3$. It is evident intuitionally that a minimum exists when the triangle has three acute angles; if it has a right or obtuse angle, say at A_1, the degenerate pedal triangle of A_1 has a less perimeter than any proper inscribed triangle.

265. Theorem. *If two triangles are inscribed in the same circle, on a common base, the line joining their orthocenters is equal and parallel to the line joining their vertices.*

For if the triangles are $A_1A_2A_3$ and $A_1'A_2A_3$, with orthocenters H and H', we have seen that

$$\overline{A_1H} = 2\,\overline{OO_1} = \overline{A_1'H'}$$

so that $A_1HH'A_1'$ is a parallelogram.

Corollary. *If four points are on a circle, forming four triangles, the orthocenters of these triangles form a figure congruent to that of the given points, with corresponding lines parallel and in opposite directions; the connectors of each of the given points with the orthocenters of the other three, are concurrent and bisect one another at a point midway between the centers of the circles.*

This remarkable figure will be the subject of further study in a later chapter (**417**). It may be noted that when four points constitute an orthocentric system, the centers of their circles constitute a congruent figure; on the other hand, when

---

* The proofs of this theorem in some of the texts are open to grave criticism. Our proof follows that of Russell, *l.c.*, p. 138.

four points are on a circle, their orthocenters constitute a congruent figure.

266. We introduce a solid figure which has interesting bearings on the properties of the orthocenter.

Theorem. *Let a semicircle be drawn on each altitude of an acute triangle as diameter, in planes perpendicular to the plane of the triangle. These circles are concurrent at a point P in the perpendicular to the plane at H. Further, a right angle is subtended at P by each side, each altitude, and by any line from a vertex to the opposite side.*

For H has equal powers with regard to the three circles on the altitudes as diameters, hence the chords of these circles perpendicular to their diameters are equal. Obviously any altitude subtends a right angle at P. To show that the same is true of a side, we compute $\overline{A_2P}^2$ and $\overline{A_3P}^2$, and find that their sum is $\overline{A_2A_3}^2$; hence A_2PA_3 is a right triangle. If then A_1P is perpendicular to A_2P and to A_3P, it is perpendicular to their plane. Since the lines PA_1, PA_2, PA_3 are mutually perpendicular, we may visualize the figure as a piece cut from a corner of a cube by an oblique plane. Conversely:

Theorem. *If three mutually perpendicular planes are cut by an oblique plane, the projection on the latter of the common point of the three planes is the orthocenter of the triangle formed.*

267. It has been proved in **91** and in **230** that the midpoints of the diagonals of a complete quadrilateral are on a line. We may now re-establish this theorem, with some further extensions.

Theorem. *The orthocenter of a triangle is a radical center for all circles, each of which passes through the extremities of an altitude; in other words, if B_1, B_2, B_3, are any points on the respective sides of a triangle, the circles on A_1B_1, A_2B_2, A_3B_3, as diameters have H as radical center.*

This is merely another restatement of **255**.

Theorem. *If three collinear points B_1, B_2, B_3 are marked on the sides of a triangle, the circles on A_1B_1, A_2B_2, A_3B_3 as diameters are coaxal.*

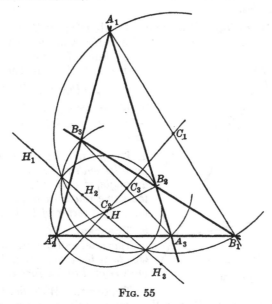

Fig. 55

For in the first place, the orthocenters of triangles $A_1A_2A_3$, $A_1B_2B_3$, $A_2B_3B_1$, $A_3B_1B_2$ do not coincide. We have noted that H, the orthocenter of $A_1A_2A_3$, is a radical center for the three circles. But consider triangle $A_1B_2B_3$; on its sides we have points B_1, A_2, A_3; and its orthocenter H' has equal power with regard to the circles on A_1B_1, B_2A_2, B_3A_3 as diameters. Thus continuing, the three circles have as radical centers the orthocenters of the four triangles, and therefore they are coaxal and the orthocenters lie on their radical axis.

268. At one volley, therefore, we have brought down the two following theorems, each a worthy prize:

Theorem of Gauss and Bodenmiller.   *The circles on the diagonals of a complete quadrilateral as diameters are coaxal.*

Theorem.   *The orthocenters of the four triangles of a complete quadrilateral are collinear on the radical axis of the aforesaid circles.*

269. Definition.   Two transversals PQ and RS to two lines APR and AQS are said to be *antiparallel* with regard to those lines if they make equal angles with them in the sense

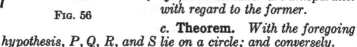

$$\angle APQ = \angle RSA;$$

in other words, if the triangles APQ and ASR are inversely similar.

a. **Theorem.**   *Two lines are antiparallel with regard to the sides of an angle, if and only if they make the same angle in opposite senses with the bisector of that angle.*

b. **Theorem.**   *If PQ and RS are antiparallel with regard to PR and QS, then the latter are antiparallel with regard to the former.*

FIG. 56

c. **Theorem.**   *With the foregoing hypothesis, P, Q, R, and S lie on a circle; and conversely.*

270. *a.* **Theorem.**   *The line joining the feet of two altitudes of a triangle is antiparallel to the third side.*

b. **Theorem.**   *The tangent to the circumcircle at a vertex is antiparallel to the opposite side.*

c. **Theorem.**   *The sides of the pedal triangle of H are parallel to the tangents to the circumcircle at the vertices.*

d. **Theorem.**   *The radius of the circumcircle at a vertex is per-*

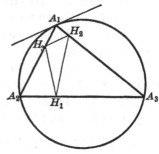

FIG. 57

pendicular to all lines antiparallel to the opposite sides; in particular, each side of triangle $H_1H_2H_3$ is perpendicular to the corresponding radius. (Cf. **250, 251.**)

271. The Median Point. The properties of the median point are not so interesting as those of the orthocenter. We will consider a few of its theorems, and will discuss briefly certain other points which are somewhat analogous to it.

272. Theorem. *Each median divides the triangle into two equal areas; all the medians together divide it into six equal parts, and the lines from the median point to the vertices divide the whole into three equivalent triangles.*

Corollary. *The perpendiculars from the median point to the sides are inversely proportional to the sides,*

$$p_1 : p_2 : p_3 = \frac{1}{a_1} : \frac{1}{a_2} : \frac{1}{a_3}$$

For $a_1p_1 = a_2p_2 = a_3p_3 = \frac{2}{3}\Delta$

273. Theorem. *The distance from M to any line is equal to one third the algebraic sum of the distances from the vertices to the same line.*

For if we denote the perpendiculars to a line XY from the points A_1, A_2, A_3, M, O_1 by d_1, d_2, d_3, d, d' respectively, we have by simple proportions

$$d = d_1 + \frac{2}{3}(d' - d_1), \quad d' = \frac{1}{2}(d_2 + d_3)$$

whence $\qquad d = \frac{1}{3}(d_1 + d_2 + d_3)$

Corollary. *If a line is so drawn that the algebraic sum of distances from the vertices of a triangle is zero, it passes through the median point. All lines for which this sum has a constant value are tangent to a circle about M.*

274. Definition. In the interest of uniformity, we define as the median point of three points A, B, C on a line that point M for which

$$\overline{MA} + \overline{MB} + \overline{MC} = O$$

In consequence of the foregoing definition, if P is any point of the line ABC,

$$\overline{PM} = \tfrac{1}{3}(\overline{PA} + \overline{PB} + \overline{PC})$$

Also M trisects the segment from any one of the given points to the point midway between the other two; and it satisfies **273**. This property represented by **273** associates the median point with the physical concept of center of gravity, and shows that the center of gravity of three equal weights at the vertices of a triangle is at M. This will be discussed further in Chapter XV.

275. Theorem. *The sum of the squares of the distances from a point to the vertices of a triangle equals three times the square of the distance from the point to the median point, plus the sum of the squares of the distances from the latter to the vertices. That is, for any point P,*

$$\overline{PA_1}^2 + \overline{PA_2}^2 + \overline{PA_3}^2 = \overline{MA_1}^2 + \overline{MA_2}^2 + \overline{MA_3}^2 + 3\,\overline{PM}^2$$

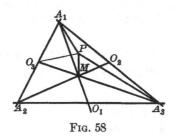

FIG. 58

We have $\overline{PA_1}^2 + \overline{PA_2}^2 = \tfrac{1}{2}a_3^2 + 2\,\overline{PO_3}^2$ **(96)**

$$2\,\overline{PO_3}^2 + \overline{PA_3}^2 = 3\,\overline{PM}^2 + 2\,\overline{O_3M}^2 + \overline{A_3M}^2 \qquad \textbf{(100)}$$

so that $\overline{PA_1}^2 + \overline{PA_2}^2 + \overline{PA_3}^2 = 3\,\overline{PM}^2 + \tfrac{1}{2}a_3^2 + \tfrac{2}{9}m_3^2 + \tfrac{4}{9}m_3^2$

Again, $m_3^2 = \tfrac{1}{4}(2\,a_1^2 + 2\,a_2^2 - a_3^2)$ and $\overline{MA_3} = \tfrac{2}{3}m_3$,

whence finally the result as stated.

Corollary. *The median point is that point in the plane for which the sum of the squares of the distances to the vertices of the triangle is minimum; and the locus of a point for which this sum has a constant value is a circle about the median point M. Also*

$$\overline{OM}^2 = R^2 - \frac{a_1^2 + a_2^2 + a_3^2}{9}$$

276. Theorem. *If the vertices of a triangle lie on the sides of another, and divide them in a fixed ratio, the triangles have the same median point.* (Cf. **107**, also **477** ff.)

Let B_1, B_2, B_3 lie on the sides of $A_1A_2A_3$, so that

$$\frac{\overline{B_1A_2}}{\overline{B_1A_3}} = \frac{\overline{B_2A_3}}{\overline{B_2A_1}} = \frac{\overline{B_3A_1}}{\overline{B_3A_2}} = \frac{m}{n}$$

We locate a point X_1 on A_2A_3, so that $\overline{X_1A_3} = \overline{A_2B_1}$, then B_2X_1 is parallel to A_1A_2, and B_3X_1 to A_1A_3; therefore A_1B_3 and B_2X_1 are equal and parallel. Now connect O_1 and P_3, the respective mid-points of B_1X_1 and B_1B_2. We have O_1P_3 parallel to X_1B_2, and equal to half of it; therefore it is parallel to B_3A_1 and equal to half of it; so that A_1O_1 and B_3P_3 trisect each other at M, as was to be shown.

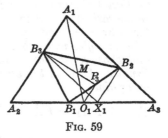

Fig. 59

This proof, which is essentially that of Fuhrmann, can be reversed, yielding the converse theorem:

Theorem. *If a triangle is inscribed in another, so that their median points coincide, the vertices of the former divide the sides of the latter in equal ratios.*

277. Exmedians. Through each vertex of a triangle there is a line, whose properties resemble closely those of the median. Such a line is called an exmedian, and we have a number of theorems concerning medians and exmedians.

Definition. The line through a vertex of a triangle, parallel to the opposite side, is called an *exmedian*. The point of intersection of two exmedians is called an *exmedian point*.

It may be noted that this is a case of the general theorem of **224.** Another illustration of the latter is furnished by the bisectors of the angles, and it will be desirable to note not only the similarity of the properties of medians and exmedians, and of the median point and exmedian points, but the resemblances between the configuration of medians and exmedians as a whole, and that of the angle bisectors.

Theorem. *Through each exmedian point passes the median from the opposite vertex. The perpendiculars from a point on an exmedian to the adjacent sides are inversely as the lengths of the sides, and the perpendiculars from an exmedian point on the three sides are inversely as the lengths of those sides. A median divides the opposite side in the ratio -1, an exmedian in the ratio $+1$. A median is divided by the median point in the ratio $-\frac{1}{2}$, and by an exmedian point in the ratio $+\frac{1}{2}$. The triangles whose bases are the sides of the given triangle, and whose vertices are at an exmedian point, are equal in area. Finally, if M' is the exmedian point opposite A_1, and P is any point,*

$$\overline{PA_2}^2 + \overline{PA_3}^2 - \overline{PA_1}^2 = \overline{PM'}^2 + \overline{M'A_2}^2 + \overline{M'A_3}^2 - \overline{M'A_1}^2$$

THE POLAR CIRCLE

278. Definition. The polar circle of a triangle is the circle whose center is the orthocenter, and whose radius is given (**255**) by

$$r^2 = \overline{HA_1} \cdot \overline{HH_1} = \overline{HA_2} \cdot \overline{HH_2} = \overline{HA_3} \cdot \overline{HH_3}$$
$$= -4R^2 \cos\alpha_1 \cdot \cos\alpha_2 \cdot \cos\alpha_3 = \tfrac{1}{2}(a_1^2 + a_2^2 + a_3^2) - 4R^2$$

It follows that the polar circle has real existence only when the triangle has an obtuse angle; and we can at once establish the following theorems for an obtuse triangle:

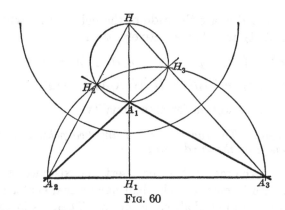

Fɪɢ. 60

Theorems. *With regard to the polar circle, each vertex and the foot of the corresponding altitude are inverse; each side is the polar of the opposite vertex. The inverse of a side is the circle having as diameter the line from the opposite vertex to the orthocenter. The circle on any side of the triangle as diameter is unchanged by the inversion, and is therefore orthogonal to the polar circle. More generally, any circle passing through a vertex and the foot of the altitude from that vertex; in other words, a circle having as diameter a line from the vertex to the opposite side, is unchanged by the inversion and is orthogonal to the polar circle. The inverse of the circumcircle with regard to the polar circle is the nine point circle (253).*

279. Theorem. *The given triangle is self-conjugate with regard to its polar circle (143); and conversely, a circle is the polar circle of every self-conjugate triangle.*

280. In an orthocentric system, three of the four triangles formed are obtuse; if, for example, $A_1A_2A_3$ is an acute triangle, H its orthocenter, the triangles A_2A_3H, A_3A_1H, A_1A_2H have real polar circles with centers respectively at A_1, A_2, A_3.

Theorem. *Any two polar circles of an orthocentric system are orthogonal.*

For if r_2 and r_3 arc the radii of the polar circles whose centers are at A_2 and A_3 respectively, then

$$r_2{}^2 = \overline{A_2H_1} \cdot \overline{A_2A_3}, \quad r_3{}^2 = \overline{A_3H_1} \cdot \overline{A_3A_2} = \overline{H_1A_3} \cdot \overline{A_2A_3}$$
$$r_2{}^2 + r_3{}^2 = (\overline{A_2H_1} + \overline{H_1A_3})\,\overline{A_2A_3} = \overline{A_2A_3}{}^2$$

which is the condition for orthogonality.

Theorem. *The radical axis of any two polar circles is the altitude from the third vertex.*

281. For the moment let us assert the existence of "imaginary circles"; such a circle shall have a real center and the square of its radius shall be negative. Then, as in the foregoing, the triangles of an orthocentric system have four polar circles, three real and one imaginary, and any two of the four are orthogonal. Conversely, if four circles are mutually orthogonal, their centers are an orthocentric system, except in the degenerate case about to be mentioned. An interesting theorem about such a system is:

Theorem. *If inversion is performed with regard successively to each of four mutually orthogonal circles, every point returns to its original position.*

For let us simplify the figure, transforming two of the four circles into mutually perpendicular straight lines. It will be seen that the other two are then concentric about the point of intersection of these lines, with radii r_2 and r_3, such that $r_2{}^2 + r_3{}^2 = 0$. Then the theorem is easily proved for this special figure, and therefore it is true for the general figure.

282. Let us consider the polar circles of the triangles determined by four lines, no two of which are perpendicular. We see that at least two of these triangles will be obtuse, and that all four may be obtuse. We consider one of the triangles, say $A'B'C'$, on the respective sides of which are three collinear points A, B, C. We have seen that the polar circle of $A'B'C'$ is orthogonal to the circles on AA', BB', CC' as diameters; hence we have additional light on the theorems of **267**.

Theorem. *The polar circles of the triangles of a complete quadrilateral constitute a coaxal system conjugate to that of the circles on the diagonals.*

283. Exercises. We close the chapter with a group of miscellaneous theorems and exercises.

a. The perpendicular bisector of H_2H_3 passes through O_1.

b. The lines from the mid-points of H_2H_3, etc., perpendicular respectively to A_2A_3, etc., are concurrent.

c. If the point of concurrence of three lines from the vertices of a triangle is also the Miquel point of the points in which the lines cut the opposite sides, then the lines are necessarily the altitudes.

d. If the altitudes of an acute triangle are extended to meet the circumcircle, the hexagon having these three chords as diagonals has twice the area of the triangle.

284. If the base A_2A_3 of a triangle, and the radius R of the circumcircle, are given, the locus of the third vertex is evidently a circle of radius R, passing through the points A_2 and A_3.

a. In this figure, what is the locus of the orthocenter: of the median point?

b. Again, if the vertex A_1 and the directions of A_1A_2 and A_1A_3 are given, as well as the length R, the locus of O is a circle of radius R with center at A_1. In this case, what is the locus of H? and so on.

c. If H_1, H_2, H_3 are given in position, then A_1, A_2, A_3 can be found. Are they uniquely determined?

d. If O_1, O_2, O_3 are given, the triangle is uniquely determined; the same is true if H_1, O_2, O_3 are given; but if three such points as H_2, H_3, O_1 are given, no solution is possible unless O_1 is equidistant from H_2 and H_3 (cf. a) in which case the triangle is indeterminate, each vertex lying on a certain circle.

285. *a. If a variable triangle is inscribed in a fixed circle on a fixed base A_2A_3, the line H_2H_3 is tangent to another fixed circle.*

For it is a chord of fixed length in the fixed circle on A_2A_3 as diameter.

b. In an orthocentric system, the median points of the four triangles also constitute an orthocentric system homothetic to the given one in the ratio $1:3$.

c. If parallel lines are drawn through the vertices of a triangle, they meet the circumcircle in points which are vertices of a triangle symmetrically congruent to the given triangle.

d. If two directly congruent triangles are inscribed in the same circle, the corresponding sides intersect in three points which are vertices of a triangle directly similar to both; the center of the circle is the common Miquel point, and also the orthocenter of the new triangle.

For if A_2A_3 meets B_2B_3 at C_1, etc., then we see easily that A_1, B_1, C_2, C_3, O are concyclic, and O is the Miquel point. But we know one Miquel triangle of O, namely its pedal triangle, which is similar to $A_1A_2A_3$ with O as its orthocenter.

e. If lines are drawn through the vertices of a triangle, making equal angles with the opposite sides, their triangle is similar to the given triangle, with its circumcenter at H.

f. If the diagonals of a simple quadrangle $ABCD$ intersect at K, the centers of the circles ABK, BCK, CDK, DAK form a parallelogram whose sides are parallel to the diagonals of the quadrangle.
Conversely, if $PQRS$ is a parallelogram, and K any point, the circles with centers at P, Q, R, S, passing through K, meet successively at four points A, B, C, D such that AC and BD meet at K.

g. If the sides of a triangle are a fixed tangent and a variable tangent to a given circle, and the chord of contact, the locus of the orthocenter is a circle equal to the given circle, with its center at the point of tangency of the fixed tangent.

For we show that the perpendicular from the circumcenter

of the triangle to the variable tangent equals half the given radius, and apply **256**.

286. The following theorems of Hagge * offer no difficulty.

Given three concurrent lines from the vertices of a triangle to the opposite sides, and circles on these as diameters; if lines through H and perpendicular to these concurrent lines meet the respective circles, the six intersections lie on a circle whose center is the point of concurrence P.

In the same figure, let the same lines through H cut the circles drawn on the sides of the triangle as diameters; the six intersections lie on a circle, whose center, with reference to triangle $A_1A_2A_3$, is homologous to the position of P in triangle $A_1A_2A_3$. The circles on the concurrent lines meet the circles on the sides as diameters in six points of a circle.

Exercise. In this chapter the following propositions are left unproved entirely or in part, and the proofs are to be completed by the reader: **250–253, 255, 262–264, 266, 269, 270, 272, 276, 277, 278, 279, 283–86.**

* *Zeitschrift für Math. und Nat. Unterricht,* 39, 1908, p. 1.

CHAPTER X

INSCRIBED AND ESCRIBED CIRCLES

287. This chapter is a study of the points of intersection of the bisectors of the angles of a triangle, and of the circles whose centers are at these points and which are tangent to the sides of the triangle.*

We are aware that the bisectors of the interior angles of a triangle are concurrent at a point I, called the *incenter*, which is equidistant from the sides of the triangle; the radius of the *incircle*, or *inscribed circle*, whose center is at the incenter and which touches the sides, shall be designated by ρ. Similarly the bisector of any interior angle, and those of the exterior angles at the other vertices, are concurrent at a point outside the triangle; these three points are called *excenters*, and the corresponding tangent circles *excircles* or *escribed circles*. The excenter lying on $A_1 I$ is denoted by J', and the radius of the escribed circle with center at J' is ρ_1. We use X_1 to denote the point where the interior bisector $A_1 I J'$ meets $A_2 A_3$, and Y_1 for the intersection of the exterior bisector $A_1 J'' J'''$ with $A_2 A_3$.

288. Some of the theorems concerning incenters and excenters can be derived at once from the results of the last chapter.

Theorem. *The incenter and excenters of a triangle are an orthocentric system; conversely, the vertices and orthocenter of a triangle are the incenter and excenters of the triangle whose vertices are the feet of the altitudes* (**253** *d*).

289. We know that $A_1 X_1$ and $A_1 Y_1$ are perpendicular to

_____

* Paragraphs 299–307 may be omitted without impairing the sequence.

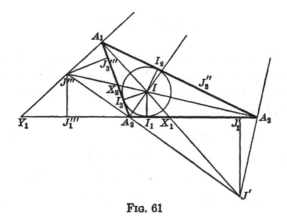

Fig. 61

each other; X_1 and Y_1 divide A_2A_3 in the ratio of $\overline{A_1A_2}$ to $\overline{A_1A_3}$. The following angle relations can be established without difficulty:

$$\angle I_2 I I_3 = 180° - \alpha_1 = \alpha_2 + \alpha_3$$

$$\angle I_2 I_1 I_3 = \tfrac{1}{2} \angle I_2 I I_3 = 90° - \frac{\alpha_1}{2} = \frac{\alpha_2 + \alpha_3}{2}$$

$$\angle A_1 I_2 I_3 = 90° - \frac{\alpha_1}{2} = \angle A_1 I_3 I_2$$

$$\angle I I_2 I_3 = \frac{\alpha_1}{2}$$

$$\angle A_1 X_1 A_2 = \alpha_3 + \frac{\alpha_1}{2} = 180° - \alpha_2 - \frac{\alpha_1}{2}$$

$$\angle I A_1 O = \frac{\alpha_2 - \alpha_3}{2}$$

$$\angle A_2 I A_3 = 90° + \frac{\alpha_1}{2}$$

with similar equations for each of the excenters, *mutatis mutandis*. In particular,

$$\angle J''J'J''' = \frac{\alpha_2 + \alpha_3}{2}$$

290. The segments on the sides of the triangle determined by the points of tangency of the inscribed and escribed circles are simply expressed in terms of the half sum of the sides, s. Namely, neglecting algebraic signs:

$$s \quad = \overline{A_1J_2}' = \overline{A_1J_3}' = \overline{A_2J_3}'' = \overline{A_2J_1}'' = \overline{A_3J_1}''' = \overline{A_3J_2}'''$$

$$s - a_1 = \overline{A_1I_2} = \overline{A_1I_3} = \overline{A_2J_3}''' = \overline{A_2J_1}''' = \overline{A_3J_1}'' = \overline{A_3J_2}''$$

$$s - a_2 = \overline{A_2I_3} = \overline{A_2I_1} = \overline{A_3J_1}' = \overline{A_3J_2}' = \overline{A_1J_2}''' = \overline{A_1J_3}'''$$

$$s - a_3 = \overline{A_3I_1} = \overline{A_3I_2} = \overline{A_1J_2}'' = \overline{A_1J_3}'' = \overline{A_2J_3}' = \overline{A_2J_1}'$$

These may be obtained algebraically, by virtue of the theorem that tangents to any circle from an external point are equal. For instance, if

$$x = \overline{A_1I_2} = \overline{A_1I_3}, \quad y = \overline{A_2I_3} = \overline{A_2I_1}, \quad z = \overline{A_3I_1} = \overline{A_3I_2}$$

we have $y + z = a_1$, $z + x = a_2$, $x + y = a_3$,
which yield, when solved simultaneously,

$$2x = a_2 + a_3 - a_1, \text{ etc.}$$

Corollaries:

$$\overline{I_1J_1}'' = a_2 = \overline{J_1'J_1}''', \quad \overline{I_1J_1}''' = a_3 = \overline{J_1'J_1}''$$

$$\overline{I_1J_1}' = a_2 - a_3, \qquad \overline{J_1''J_1}''' = a_2 + a_3,$$

$$\overline{O_1I_1} = \tfrac{1}{2}(a_2 - a_3), \qquad \overline{O_1J_1}' = \tfrac{1}{2}(a_2 + a_3)$$

291. Geometric corollaries:

a. The lines from the vertices to the points of contact of the inscribed circle meet in a point (Gergonne point).

b. The lines from the vertices to the internal points of contact of the respective escribed circles meet at a point (Nagel point, see 361).

c. The points of Gergonne and Nagel are isotomic conjugates.

d. More generally, the lines connecting the vertices with the points of contact of the inscribed and escribed circles are concurrent three by three at eight points, four pairs of isotomic conjugates.

e. The perpendiculars to the sides of the triangle at these same points of contact are concurrent at the in- and excenters, and also at four other points.

These new points are the circumcenters of the orthocentric system of the in- and excenters.

f. If B_1, B_2, B_3 are the excenters of triangle $A_1A_2A_3$, $C_1C_2C_3$ those of $B_1B_2B_3$, and so on; show that the triangles of this series tend to become more and more nearly equilateral.

(Express each angle as a sum $60° \pm x$, and apply the last equation of **289**).

292. We designate by P_1 the point where the bisector A_1I meets the circumcircle, namely the mid-point of arc A_2A_3; and by Q_1 the diametrically opposite point where the exterior bisector $A_1J''J'''$ meets the circle.

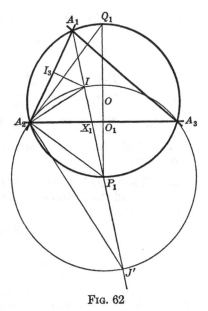

Theorem. *The circle on IJ' as diameter passes through A_2 and A_3; its center is at P_1, and its radius is*

$$r = \frac{a_1}{2 \cos \dfrac{\alpha_1}{2}} = 2R \sin \frac{\alpha_1}{2}$$

For evidently $IA_2 J'$ and $IA_3 J'$ are right angles, and the circle on IJ' as diameter passes through A_2 and A_3. But the center of the circle,

Fig. 62

being on IJ', is also on the perpendicular bisector of A_2A_3, and therefore is at P_1. The formulas for the radius are immediate.

Theorem. *The circle on $J''J'''$ as diameter passes through A_2 and A_3; its center is at Q_1, and its radius*

$$r' = \frac{a_1}{sin \frac{\alpha_1}{2}} = 2R \, cos \, \frac{\alpha_1}{2}$$

Corollary. *These two circles are orthogonal.*

293. It will be noted that these circles are the same as were discussed in **251**, with reference to the orthocentric system $J'J''J'''I$. Numerous consequences are evident.

a. $$\overline{IJ'} = a_1/cos \, \frac{\alpha_1}{2} = 4 \, R \, sin \, \frac{\alpha_1}{2}$$

b. $$\overline{A_3I} = \overline{IJ} \, sin \, A_3A_2I = 4 \, R \, sin \, \frac{\alpha_1}{2} \, sin \, \frac{\alpha_2}{2},$$

$$\rho = 4 \, R \, sin \, \frac{\alpha_1}{2} \, sin \, \frac{\alpha_2}{2} \, sin \, \frac{\alpha_3}{2}$$

c. $$\overline{P_1I}^2 = \overline{P_1X_1} \cdot \overline{P_1A_1}, \quad \overline{O_1I_1}^2 = \overline{O_1X_1} \cdot \overline{O_1H_1}$$

d. $$\overline{X_1I_1} \cdot \overline{X_1J_1'} = \overline{X_1O_1} \cdot \overline{X_1H_1}$$

For $$\overline{X_1I} \cdot \overline{X_1J'} = \overline{X_1A_3} \cdot \overline{X_1A_2} = \overline{X_1A_1} \cdot \overline{X_1P_1}$$

294. Problem.   *To construct a triangle, given O, I, J' in position.*

295. Theorem.* *The radii of the circumscribed and inscribed circles, and the distance d between their centers, are connected by the equation*

$$R^2 - d^2 = 2 \, R\rho$$

or $$\frac{1}{R-d} + \frac{1}{R+d} = \frac{1}{\rho}$$

---

* For the early history of this important theorem, see Mackay, *Proceedings of Edinburgh Math. Society*, V, 1886–87, p. 62.

Either of these equations can be derived algebraically from the other; we shall establish the first. We see at once (figure 62) that triangles $Q_1P_1A_2$ and A_1II_3 are similar, hence:

$$\overline{Q_1P_1} \cdot \overline{II_3} = \overline{P_1A_2} \cdot \overline{A_1I}$$

which gives at once

$$2R\rho = \overline{A_1I} \cdot \overline{P_1I} = R^2 - d^2$$

Similarly, if d_1 is the distance from O to the excenter J',

$$2R\rho_1 = d_1^2 - R^2, \qquad \frac{1}{R - d_1} + \frac{1}{R + d_1} = \frac{1}{\rho_1}$$

By suitable conventions as to signs, these two propositions may be regarded as equivalent forms of the same theorem. The converse theorem, that if two circles satisfy an equation of this form, then a triangle may be inscribed in the one and circumscribed to the other, may be established with a little difficulty from the same figure, but we propose to use an alternative method based on inversion.

296. Theorem. *If an inversion is performed with regard to the incircle of a triangle, the sides and the circumcircle are transformed into equal circles of diameter ρ; the inverse of A_2A_3 is the circle on II_1 as diameter, and that of the circumcircle is the circle through the mid-points of the sides of triangle $I_1I_2I_3$.*

For A_1I_2 and A_1I_3 are tangent to the incircle, therefore the inverse of A_1 is the mid-point of I_2I_3. It may be noted that the circle through these mid-points is the nine point circle (**258**) of triangle $I_1I_2I_3$ (see also **104** *a*). By virtue of the formula for the radius of a circle inverse to a given circle (**71**), the converse of **295** is a corollary of this theorem.

297. Theorem. *If two circles, O (R) and $I(\rho)$, are so situated that*

$$R^2 - \overline{OI}^2 = 2R\rho$$

*then a triangle can be drawn with its vertices on the first circle
and its sides tangent to the second; and that in an infinite
number of ways, with any point of the first circle as a vertex.*

We effect an inversion with regard to the second circle, and
let the first be transformed into a circle with center O' and
radius R'. Then by **71**,

$$R' = \frac{\rho^2}{\overline{OI}^2 - R^2}\, R$$

whence by hypothesis, $R' = \frac{1}{2}\rho$. Let A' be any point on this
new circle; let the circles of diameter ρ, passing through I and
A', cut the circle at B' and C' and be tangent to the circle I
(ρ) at Z and Y. Then we show easily that A', B', C' are
mid-points of the sides of a triangle XYZ inscribed in the
circle I (ρ); and that I, A', B', C' constitute an orthocentric
system. Reverting to the original figure, we have three lines
tangent to the circle I (ρ) at X, Y, Z, and intersecting at
three points A, B, C on the original circle O (R).

Corollary. *If two circles admit a triangle inscribed to one
and circumscribed to another, they admit an infinite number.*

The similarity of the equations given above to that of **125**
should be noted.

In a problem of construction, if two of the lengths R, ρ, $\overline{OI}$
are given, we may use the above relation to find the third,
and draw the two circles. Then the triangle can usually be
constructed. Again, the equation shows that the radius of
the incircle is always less than half that of the circumcircle,
except in an equilateral triangle.

Theorem. *If XY is a diameter of the incircle perpendicular
to OI, the perimeter of triangle OXY is equal to the diameter
of the circumcircle.*

298. Many of the relations subsisting among the various
parts of the triangle may best be expressed by algebraic

equations and formulas. This is especially true of the radii of the inscribed and escribed circles. We have already noted a number of such equations of simple type; and it will be of interest at this time to enumerate some others whose derivation is not especially difficult.*

a. $$\rho = \frac{\Delta}{s} = 4\,R\,sin\,\frac{\alpha_1}{2}\,sin\,\frac{\alpha_2}{2}\,sin\,\frac{\alpha_3}{2} \qquad (\textbf{15}\ d,\ \textbf{293}\ b)$$

$$\rho_1 = \frac{\Delta}{s - a_1} = 4\,R\,sin\,\frac{\alpha_1}{2}\,cos\,\frac{\alpha_2}{2}\,cos\,\frac{\alpha_3}{2}$$

b. $$\frac{1}{\rho_1} + \frac{1}{\rho_2} + \frac{1}{\rho_3} - \frac{1}{\rho} = 0$$

c. $$\rho_1 + \rho_2 + \rho_3 = 4\,R + \rho$$

For $\rho_1 + \rho_2 + \rho_3 - \rho$ reduces to $\dfrac{a_1 a_2 a_3\,\Delta}{s(s - a_1)(s - a_2)(s - a_3)}$

$$= \frac{a_1 a_2 a_3}{\Delta} = 4\,R$$

d. $$R\rho = \frac{a_1 a_2 a_3}{4\,s}$$

e. $$\frac{1}{h_1} + \frac{1}{h_2} + \frac{1}{h_3} = \frac{1}{\rho} = \frac{1}{\rho_1} + \frac{1}{\rho_2} + \frac{1}{\rho_3},$$

$$\frac{1}{\rho_1} = \frac{1}{h_2} + \frac{1}{h_3} - \frac{1}{h_1}, \qquad \frac{1}{\rho_2} + \frac{1}{\rho_3} = \frac{1}{\rho} - \frac{1}{\rho_1} = \frac{2}{h_1}$$

For $\dfrac{1}{h_1} = \dfrac{a_1}{2\,\Delta}$, etc.

---

* The reader who wishes to pursue this subject further will be amply gratified by two papers of Mackay, "Formulas Connected with the Radii of the Incircle and the Excircles of a Triangle," *Proceedings of Edinburgh Math. Society,* 12, pp. 86–105, 13, 103–04; "Properties Connected with the Angular Bisectors of a Triangle," *ibid.,* 13, pp. 37–102. Both these papers contain numerous verbal theorems, but both consist largely of formulas of the sort given in the text; the first contains some fifteen solid pages of such formulas, the second over twenty-five. A similar list, but not so well constructed, is a little book by Schroeder (*Das Dreieck und seine Beruhrungskreise*). Marcus Baker (*Annals of Mathematics,* I, p. 134) gives a list of 110 formulas for the area of a triangle.

f. $\overline{OO_1} + \overline{OO_2} + \overline{OO_3} = R + \rho$

For applying Ptolemy's theorem to $OO_1A_3O_2$, etc.,

$$\frac{a_3}{2}R = \overline{OO_2} \cdot \frac{a_1}{2} + \overline{OO_1} \cdot \frac{a_2}{2}; \text{ etc.}$$

Also $s\rho = \frac{1}{2}(\overline{OO_1} \cdot a_1 + \overline{OO_2} \cdot a_2 + \overline{OO_3} \cdot a_3)$

Adding, we may divide out a common factor s.

g. $\Delta^2 = \rho\rho_1\rho_2\rho_3$

h. $\overline{OI}^2 + \overline{OJ'}^2 + \overline{OJ''}^2 + \overline{OJ'''}^2 = 12R^2$ (295)

i. The power of I with regard to the circumcircle is $\dfrac{a_1a_2a_3}{a_1 + a_2 + a_3}$

299. In 1822, Karl Wilhelm Feuerbach (1800–1834), a teacher in the Gymnasium at Erlangen, Germany, published a small book * containing a remarkable collection of theorems on the triangle. The chief fame of this work rests on the well-known theorem bearing the name of the author, but even without this theorem, it would have been an exceedingly valuable contribution to the geometry of the triangle. Indeed, there is nothing to indicate that the author was especially impressed with the theorem alluded to, or that he regarded it as any more important than the rest of his work.

The book consists mainly of proportions and other algebraic relations among the various dimensions of the triangle, especially among the distances associated with the circumcenter, orthocenter, incenter, and excenters. Many of the results which we have already given were included by him; we quote a few others of his most striking formulas.

a. $\rho_2\rho_3 + \rho_3\rho_1 + \rho_1\rho_3 = s^2$

   $\rho(\rho_2\rho_3 + \rho_3\rho_1 + \rho_1\rho_2) = s\Delta = \rho_1\rho_2\rho_3$

   $\rho(\rho_1 + \rho_2 + \rho_3) = a_2a_3 + a_3a_1 + a_1a_2 - s^2$

* *Eigenschaften einiger merkwurdigen Punkte des geradlinigen Dreiecks, und mehrerer durch sie bestimmten Linien und Figuren.* (Nürnberg, 1822.)

$$\rho\rho_1 + \rho\rho_2 + \rho\rho_3 + \rho_1\rho_2 + \rho_2\rho_3 + \rho_3\rho_1 = a_2a_3 + a_3a_1 + a_1a_2$$

$$\rho_2\rho_3 + \rho_3\rho_1 + \rho_1\rho_2 - \rho\rho_1 - \rho\rho_2 - \rho\rho_3 = \tfrac{1}{2}(a_1^2 + a_2^2 + a_3^2)$$

These are easily proved by transformation of **298** a.

b. *The perimeter of triangle* $H_1H_2H_3$ *is* $2\,\Delta/R$.

For $\qquad \rho = a_1 \cos \alpha_1 + a_2 \cos \alpha_2 + a_3 \cos \alpha_3$

$$= a_1^2 \frac{a_2^2 + a_3^2 - a_1^2}{2\,a_1a_2a_3} + \text{etc.} = \frac{8\,\Delta^2}{a_1a_2a_3} = \frac{2\,\Delta}{R}$$

c. *The distance between the feet of the perpendiculars from* H_1 *to* A_1A_3 *and* A_1A_2 *equals half the perimeter* p.

d. *The product of the three altitudes is* $p\Delta$.

e. $a_1^2 + a_2^2 + a_3^2 + \overline{A_1H}^2 + \overline{A_2H}^2 + \overline{A_3H}^2 = 12\,R^2$ (Cf. **252** f)

f. $\overline{A_1H} + \overline{A_2H} + \overline{A_3H} = 2\,\rho + 2\,R$ $\qquad\qquad$ (**298** f)

or $\cos\alpha_1 + \cos\alpha_2 + \cos\alpha_3 = 1 + \dfrac{\rho}{R}$

g. By introducing the radius r of the circle inscribed in the triangle $H_1H_2H_3$, Feuerbach establishes a number of remarkably simple formulas (see also **324**):

$$r = \overline{HH_1}\, \cos\alpha_1 = 2\,R\, \cos\alpha_1 \cos\alpha_2 \cos\alpha_3$$

$$\overline{A_1H} \cdot \overline{HH_1} = 2\,Rr$$

$$\frac{\text{area } H_1H_2H_3}{\Delta} = \frac{r}{R}$$

$$a_1^2 + a_2^2 + a_3^2 = 4\,rR + 8\,R^2 \qquad\qquad \text{(Cf. } \mathbf{255}\text{)}$$

h. Finally combining the last equation with that of *e*,

$$\overline{A_1H}^2 + \overline{A_2H}^2 + \overline{A_3H}^2 = 4\,R^2 - 4\,Rr$$

300. The principle of transformation. In such developments as those of the last few pages, a theorem about the inscribed circle of a triangle suggests an analogous theorem about each excircle, and vice versa. In some cases we have stated and proved the related theorem; but the precise method of formulation is not clear except in the simplest cases.   This problem has been the subject of considerable study, and a set of general rules for transforming equations has been established.   We will, without discussing the subject in detail, briefly state these rules.*

Denoting by l_1, l_2, l_3 the lengths of the bisectors of the interior angles, by λ_1, λ_2, λ_3 those of the exterior angles; let all other letters have their usual meaning.   If then we make the following substitutions in any triangle formula, we obtain a valid formula.

Replace	a_1	a_2	a_3	s	$s-a_1$	$s-a_2$	$s-a_3$
by	a_1	$-a_2$	$-a_3$	$-(s-a_1)$	$-s$	$s-a_3$	$s-a_2$

Replace	ρ	ρ_1	ρ_2	ρ_3	R
by	ρ_1	ρ	$-\rho_3$	$-\rho_2$	$-R$

Replace	h_1	h_2	h_3	α_1	α_2	α_3	Δ
by	$-h_1$	h_2	h_3	$-\alpha_1$	$180-\alpha_2$	$180-\alpha_3$	$-\Delta$

Replace	l_1	l_2	l_3	λ_1	λ_2	λ_3
by	$-l_1$	$-\lambda_2$	$-\lambda_3$	$-\lambda_1$	$-l_2$	$-l_3$

Quantities not listed may be similarly accounted for.   This scheme should be verified by the reader, by experimenting with the formulas of the preceding sections, and elsewhere.

301. Theorem.   *In a triangle, the outer common tangents to the excircles form a triangle whose incenter coincides with the circumcenter of triangle $J'J''J'''$, and the radius of whose incircle is*

$$r = 2R + \rho = \tfrac{1}{2}(\rho + \rho_1 + \rho_2 + \rho_3) \qquad \text{(Cf. \textbf{298} c)}$$

We omit the proof, which is long and dull.   The theorem is noteworthy as being equivalent to a theorem in a Japanese geometry of about 1820; it was more recently rediscovered in Europe.†

---

* Mackay, *Proceedings of Edinburgh Math. Society*, XII, p. 87; Lemoine, *Bulletin Soc. Math. de France*, XIX, p. 133; *Proceedings of Edinburgh Math. Society*, XIII, p. 2; Lucas, *Nouvelles Correspondances Math.*, II, p. 384; *ibid.*, III, p. 1.

† *Mathesis*, 1896, p. 192; 1898, p. 203; 1911, p. 208.

Exercise.   Apply the principle of transformation (**300**) to this theorem.

302. Another theorem of Oriental origin may be noted here.   According to T. Hayashi, it was the ancient custom of Japanese mathematicians to inscribe their discoveries on tablets which were hung in the temples, to the glory of the gods and the honor of the authors.   The following theorem is known to have been thus exhibited in 1800.*

Theorem. *Let a convex polygon, inscribed in a circle, be divided into triangles by diagonals from one vertex.   The sum of the radii of the circles inscribed in these triangles is. the same, whichever vertex is chosen.*

It is evident that if the theorem can be proved for a quadrangle, it can be proved by induction for any polygon.   A proof for the quadrangle can be based on **298** f.

303. Theorem. (Feuerbach).   *If the incenters of triangles $A_1H_2H_3$, $A_2H_3H_1$, $A_3H_1H_2$ are X_1, X_2, X_3 respectively, then X_2X_3 is equal and parallel to I_2I_3; X_1, X_2, X_3 are the reflections of I with regard to the sides of triangle $I_1I_2I_3$.*

For let X_2Y be perpendicular to A_2A_3; then, since we know that triangles $A_2A_1A_3$ and $A_2H_1H_3$ are similar, in the ratio $1 : \cos \alpha_2$, and that I and X_2 are homologous points, and also I_3 and Y;

$$\frac{\overline{A_2Y}}{\overline{A_2I_3}} = \cos \alpha_2$$

It follows that I_3Y is perpendicular to A_2A_3, and therefore passes through X_2.   In other words, I_2X_3, I_3X_2, II_1 are perpendicular to A_2A_3; and similarly for the other sides.   It is then easy to establish the fact that $II_1 X_2 I_3$, for example, is a rhombus.

304. Theorem. *The chords joining points of contact of the*

* *Mathesis*, 1906, p. 257.

inscribed circle are parallel to the corresponding exterior bisectors. Similarly, the chords of contact of an escribed circle are parallel respectively to the exterior bisector of the opposite angle and the interior bisectors of the other two angles.

Corollary.   *The pedal triangle of the incenter or an excenter is similar and homothetic to the triangle of the other three points.*

305. Theorem.   *Triangles $P_1P_2P_3$ (292) and $J'J''J'''$ are homothetic, with I as center of similitude, ratio $2:1$.*
The incenter I is the orthocenter of $P_1P_2P_3$.
Triangles $P_1P_2P_3$ and $I_1I_2I_3$ are also homothetic.

306.   *The radical axes of the in- and excircles are the bisectors of the angles of triangle $O_1O_2O_3$.*

307. Exercises.   Some miscellaneous exercises will conclude the chapter.

a. If AB and AC are fixed lines, and XY any transversal, the bisectors of angles AXY and AYX meet in a point P, whose locus consists of the bisectors of the angle BAC.

b. If in triangle $A_1A_2A_3$, A_1M and A_1N are perpendicular to A_2I and A_3I, MN is parallel to A_2A_3.

c. If a triangle of given perimeter has one fixed angle, the third side touches a fixed circle.

d. If the sides of a triangle are three of the common tangents to two given circles, the circumcircle of the triangle passes through the point midway between the centers of the circles. The four circumcircles of a complete quadrilateral composed of the four tangents to two circles, are concurrent at the same point.

Exercise.   Complete the proofs of the following: **288, 289, 290, 291, 292, 293, 294, 297, 298, 299, 302–307.**

CHAPTER XI

THE NINE POINT CIRCLE

308. We continue the study of the triangle with a consideration of the so-called Nine Point Circle briefly mentioned in **258**, and after a survey of its more elementary properties we discuss at length the famous theorem of Feuerbach. Let us first state again the definitive theorem for the circle.

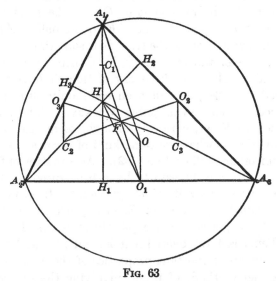

Fig. 63

Theorem. *The circle whose center F is midway between the circumcenter and the orthocenter, and whose radius is half that of the circumcircle, passes through nine notable points, namely the feet of the altitudes, the mid-points of the sides, and the midpoints of the segments from the orthocenter to the vertices.*

Denoting the mid-points of A_1H, A_2H, A_3H by C_1, C_2, C_3

respectively, we wish to prove $C_1, C_2, C_3, H_1, H_2, H_3, O_1, O_2, O_3$ lie on a circle whose center is the point F midway between O and H, and whose radius is $\frac{1}{2} R$. We can establish the existence and the properties of this circle in numerous ways; perhaps the following is the most elementary.

First, $O_2O_3C_2C_3$ is a rectangle, and therefore O_2C_2 and O_3C_3 are equal and bisect each other; say at X. Then X is the center of a circle through $O_1, C_1, O_2, C_2, O_3, C_3$. But $C_1H_1O_1$ is a right angle, and therefore H_1 lies on the circle as well; and similarly for H_2 and H_3. The center X therefore lies on the perpendicular bisector of O_1H_1, which bisects OH, and X is therefore at F, the mid-point of OH. The radii C_1F and A_1O are evidently parallel, and C_1F equals half of A_1O.

The nine point circle of the triangle is sometimes designated as Euler's circle, and frequently, by Continental writers, as Feuerbach's. That the discovery of this circle is not to be attributed to Euler, as had been commonly supposed, was established by the indefatigable Mackay.[*] The erroneous imputation to Euler, curiously enough, seems to have been the result of somewhat the same kind of error as in the case of the Simson or Wallace line (**192**). As a matter of fact, the theorem can hardly be said to have been discovered at any one time; apparently it "just growed." It is implied in problems which appeared in English journals in 1804 and 1807; and was perhaps first explicitly stated by Poncelet, in 1821. Feuerbach discovered it independently and published it, with new and important properties of the circle, in 1822.[†]

309. Other methods of proof, involving the results established in the preceding chapters, will bring out further properties of the circle.

[*] "History of the Nine-point Circle," *Proceedings of Edinburgh Math. Society*, XI, 1892, p. 19.

[†] For detailed information as to the history of the theorem, cf. Mackay, *l.c.*; Simon, *l.c.*, pp. 125–30; J. Lange, *Geschichte des Feuerbach'schen Kreises*, Berlin, 1894.

Since O and H are isogonal conjugates they have, as pointed out in **258**, a common pedal circle whose center is midway between them.   That is, the pedal circle of the orthocenter passes through the mid-points of the sides.   But in triangle A_2A_3H, the orthocenter is A_1, its pedal triangle is $H_1H_2H_3$, and therefore the circle through H_1, H_2, H_3 bisects A_2H and A_3H.   Thus:

Theorem.   *The four triangles of an orthocentric system have a common nine point circle.*

310. Another approach is by way of similar figures.   If the lines from the orthocenter H to each of the nine points in question are extended to double length, we easily see (**254**, **260**) that the extremities lie on the circumcircle.   Hence

Theorem.   *The external center of similitude of the circumcircle and the nine point circle is the orthocenter.   In other words, the nine point circle bisects any line from the orthocenter to a point on the circumcircle.*

Theorem.   *The internal center of similitude is the median point M.*

This is obvious, since the internal center must trisect FO. Again, we recognize M as the homothetic center of the similar triangles $A_1A_2A_3$ and $O_1O_2O_3$, and the circumcircle of the latter is the nine point circle under discussion.

311. We have shown that it is proper to speak of the nine point circle of an orthocentric system.

Theorem.   *The nine point circle of the incenter and excenters of a triangle is the circumcircle.*

Theorem.   *The circumcircle of a triangle bisects each of the connectors of the incenter and the excenters (**292**).*

Theorem.   *The sum of the powers of the vertices with regard to the nine point circle is $\frac{1}{4}(a_1^2 + a_2^2 + a_3^2)$.*

For the power of each vertex may be expressed by either
of two formulas, and therefore by half their sum,

$$\text{Power of } A_1 = \tfrac{1}{4}(a_2 \cdot \overline{A_1 H_3} + a_3 \cdot \overline{A_1 H_2})$$

Adding these three expressions, we obtain the result given.

Corollary. $\overline{FA_1}^2 + \overline{FA_2}^2 + \overline{FA_3}^2 + \overline{FH}^2 = 3 R^2$ **(298 h)**

312. Theorem. *All triangles inscribed in a given circle, and
having a given point as orthocenter, have the same nine point
circle.*

313. Problem. *To construct a triangle, given the circumcircle, a point A_1 on it, and the orthocenter H.*

Discuss completely, and ascertain the conditions under
which there will be a solution.

314. It is interesting to note the large number of triangles
closely related to the given triangle, which all have the same
nine point circle.   For example, if a triangle is so drawn that
the feet of its altitudes are at the points O_1, O_2, O_3, then its
nine point circle is the same as that of the given triangle.
Such a triangle may be constructed by taking as sides three
of the bisectors of the angles of $O_1 O_2 O_3$.   The new triangle
may in turn be replaced by a third, and so on; we get an unlimited sequence of triangles having a common nine point
circle, and incidentally, equal circumcircles.   If, for instance,
we repeatedly take as sides of the new triangle the exterior
bisectors of the angles, the triangles of the sequence tend more
and more nearly to the form of an equilateral triangle **(291 f)**
circumscribed to the fixed circle.

Again, we may take the feet of the altitudes of any triangle
as mid-points of the sides of a new triangle having the same
nine point circle.   We may, moreover, use the triangle $C_1 C_2 C_3$
in the same way; and by combining these various devices in
any order, we see that there are an unlimited number of
associated triangles having the same nine point circle.

315. We have an interesting coaxal system of circles including the circumcircle, the nine point circle, and the polar circle.

Theorem. *The lines connecting the feet of the altitudes meet the opposite sides in collinear points.*

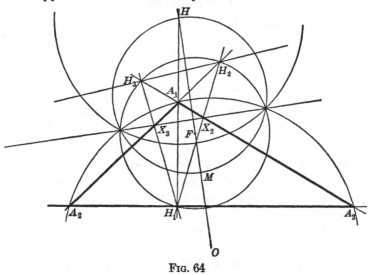

FIG. 64

This is a special case of **223**; the line, being the trilinear polar of the orthocenter, is called the *polar axis* of the triangle. If A_2A_3 meets H_2H_3 at X_1, then

$$\overline{X_1H_2} \cdot \overline{X_1H_3} = \overline{X_1A_2} \cdot \overline{X_1A_3}.$$

The polar axis of a right triangle is defined as the tangent to the circumcircle at the vertex of the right angle; let the reader show that this is consistent with the general case.

316. Theorem. *The polar axis is the radical axis of the circumcircle and the nine point circle $H_1H_2H_3$; it is therefore perpendicular to the Euler line OH. The coaxal system is of the first, second or third type, according as the triangle is acute, obtuse, or right.*

317. Theorem. *The circle on HM as diameter is a member* *of the coaxal system; it is the circle of similitude of the circum-circle and the nine point circle* (**115**).

318. Theorem. *The polar circle (if it exists) is a member of the same system, being a circle of antisimilitude* (**126**) *for the circumcircle and the nine point circle.*

319. Theorem. *The circle through the points of intersection of the tangents to the circumcircle at A_1, A_2, A_3 is a member of the coaxal system, being inverse to the nine point circle with regard to the circumcircle.*

THE THEOREM OF FEUERBACH

320. Theorem. *The nine point circle of a triangle is tangent to the inscribed circle and to each of the escribed circles.*

This is perhaps the most famous of all theorems of the triangle, aside from those known in ancient times. We have already mentioned that it was first stated and proved by Feuerbach in his classic memoir, in 1822. A few years later, it was discovered independently and stated without proof, by Steiner; and in the last hundred years it has many times been rediscovered.

Of the numerous proofs which have been contributed to the history of this theorem, none is really simple. Though all proofs are necessarily based on the same underlying principles, there are several essentially distinct methods of approach. We shall consider a few typical proofs, each of which will furnish a somewhat different aspect of the theorem, and will add to our appreciation of its beauty.

321. A proof which in itself is simple and direct can be based on the rather difficult theorem which was established in the optional portion of Chapter V, in **117**. The theorem in question was obviously devised in the first place for this present purpose. Paraphrased to fit the present problem, it states that the circle through three points O_1, O_2, O_3 will be tangent to a given circle I (ρ) if the tangents to the latter from the points, namely $\overline{O_1I_1}$, $\overline{O_2I_2}$, $\overline{O_3I_3}$, and the distances among the three points themselves, satisfy the equation

$$\overline{O_1I_1}\cdot\overline{O_2O_3} + \overline{O_2I_2}\cdot\overline{O_3O_1} + \overline{O_3I_3}\cdot\overline{O_1O_2} = 0$$

Now by **290** $\overline{O_1I_1} = \tfrac{1}{2}(a_2 - a_3)$, etc.,

and $\overline{O_2O_3} = \tfrac{1}{2}a_1$, etc.

Substituting these values we have at once an identity; thus proving that the incircle is tangent to the nine point circle. By a slight modification we can establish a proof for the excircles.

322. The following proof is designed to be as simple as possible, in the sense of demanding a minimum of prerequisite knowledge. The procedure consists of determining a point common to the circles in question, and showing that they have there a common tangent. At best, the details will not be easy; they are worked out in the following sequence of lemmas.

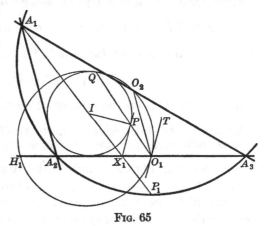

FIG. 65

Lemma 1. *Assuming that $\alpha_2 > \alpha_3$, the line O_1T, tangent to the nine point circle at O_1, makes the angle $\alpha_2 - \alpha_3$ with the base A_2A_3.*

For $\angle A_3O_1T = \angle A_3O_1O_2 - \angle O_2O_1T$
$$= \angle A_3A_2A_1 + \angle O_2O_3O_1 = \alpha_2 - \alpha_3$$

Lemma 2. *If X_1 denotes, as before, the point of intersection of the base A_2A_3 and the angle bisector A_1I, and if from X_1 we have X_1P tangent to the incircle at P and X_1P' tangent to the excircle J' at P' then PX_1P' is a straight line parallel to O_1T.*

For the transverse common tangents to the incircle and excircle meet at X_1 on their line of centers;

$$\angle A_1 X_1 P = \angle A_2 X_1 A_1 = \frac{\alpha_1}{2} + \alpha_3$$

$$\angle P X_1 A_3 = 180° - \angle A_2 X_1 A_1 - \angle A_1 X_1 P = \alpha_2 - \alpha_3$$

Lemma 3. *If the line O_1P meets the incircle again at Q, then Q is also on the nine point circle.*

For $$\overline{O_1P} \cdot \overline{O_1Q} = \overline{O_1I_1}^2 = \overline{O_1X_1} \cdot \overline{O_1H_1} \qquad \textbf{(293 c)}$$

Thus P, Q, X_1, H_1 are concyclic, and we have

$$\measuredangle O_1QH_1 = \measuredangle PX_1O_1 = \alpha_2 - \alpha_3 = \measuredangle TO_1, A_2A_3$$

Therefore, since O_1T is a tangent to the nine point circle, and

$$\measuredangle O_1QH_1 = \measuredangle TO_1H_1$$

it follows that Q is on the nine point circle.

Lemma 4. *The two circles intersecting at Q, the incircle and the nine point circle, have there the same tangent.*

For two tangents to a circle make equal angles with the chord of contact. Since the tangent to the incircle at P and the tangent to the nine point circle at O are parallel, the tangents to the two circles at Q make the same angle with O_1PQ, and therefore coincide.

Thus we have not only established the theorem that the incircle and the nine point circle are tangent, but we have a simple construction for the point of tangency. As before, the proof can be applied with slight changes to the excircles.

This proof resembles somewhat the earliest purely geo-

metric proof of the theorem, published by J. Mention * in
1850, previous proofs having been based on algebraic methods.   Mention's proof includes some of the foregoing principles, and some of those utilized in the next proof.

323. We now outline a proof which establishes the theorem
by inversion for any two of the tangent circles simultaneously.   For the
sake of definite-
ness, we will con-
sider the two es-
cribed circles with
centers at J'' and
J'''.   Three of the
common tangents
to these two cir-
cles are the sides
of the triangle.
We   locate   the

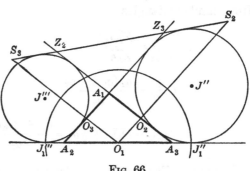

Fig. 66

fourth; then we take as circle of inversion the circle with
center at O_1, orthogonal to the two escribed circles, and show
by a computation which is simpler than it looks, that the
inverse of the nine point circle with regard to this circle is
the fourth common tangent.   Since the escribed circles are
unaltered by the inversion, this completes the proof.

Lemma 1. *The circle with center O_1, radius $\frac{1}{2}(a_2 + a_3)$, is
orthogonal to the escribed circles at J_1'' and J_1'''.*

Lemma 2. *The fourth common tangent to the escribed circles
cuts A_1A_3 at Z_2, and A_1A_2 at Z_3, so that*

$$\overline{Z_2A_1} = \overline{A_1A_2}, \quad \overline{Z_3A_1} = \overline{A_1A_3}$$

For the exterior angle bisector $J''A_1J'''$ is an axis of sym-

* *Nouvelles Annales de Math.*, 9, p. 401.

metry for the four common tangents, and A_2A_3 is the given direct common tangent.

Lemma 3. *If O_1O_2 and O_1O_3 are extended to meet Z_2Z_3 at S_2 and S_3, then*

$$\overline{O_1O_2}\cdot\overline{O_1S_2} = \overline{O_1O_3}\cdot\overline{O_1S_3} = \tfrac{1}{4}(a_2 + a_3)^2$$

For, recognizing that $O_1O_2S_2$ is parallel to $A_2A_1Z_3$, we have from similar triangles

$$\frac{\overline{O_2S_2}}{\overline{A_1Z_3}} = \frac{\overline{Z_2O_2}}{\overline{Z_2A_1}}.$$

or

$$\overline{O_2S_2} = \frac{a_2\left(a_3 + \tfrac{1}{2}a_2\right)}{a_3}$$

$$\overline{O_1S_2} = \overline{O_2S_2} + \tfrac{1}{2}a_3 = \frac{a_2{}^2 + 2\,a_2a_3}{2\,a_3} + \frac{a_3}{2} = \frac{(a_2 + a_3)^2}{2\,a_3}$$

Lemma 4. *The inverse of the tangent line $S_2Z_2Z_3S_3$ with regard to the circle $O_1(O_1J_1'')$ is the circle through O_1,O_2,O_3; that is, the nine point circle.*

For the radius of inversion is $\dfrac{a_2 + a_3}{2}$, by **290**; and thus O_2, S_2 and O_3, S_3 are pairs of inverse points.

Lemma 5. *Since the line $S_2Z_2Z_3S_3$ is tangent to the excircles, and these are unchanged by the inversion, the inverse of the line, the nine point circle, is also tangent to the excircles.*

324. Another type of proof consists in computing the distance between the center of the nine point circle and the incenter or an excenter, and showing that it actually equals the sum or the difference of the corresponding radii. This was the process used by Feuerbach, whose treatment of the subject, as we have said, was mainly algebraic. The proof of Feuerbach is based on the following steps, each of which he establishes by main strength.   Denoting by r the radius of the circle inscribed in the pedal triangle $H_1H_2H_3$,

$$\overline{OI}^2 = R^2 - 2R\rho \qquad\qquad \text{(Cf. \textbf{295})}$$
$$\overline{IH}^2 = 2\rho^2 - 2Rr$$
$$\overline{OH}^2 = R^2 - 4Rr$$
$$\overline{FI}^2 = \tfrac{1}{2}(\overline{OI}^2 + \overline{HI}^2) - \overline{FH}^2$$
$$= \tfrac{1}{4}R^2 - R\rho + \rho^2 = (\tfrac{1}{2}R - \rho)^2$$

as was to be proved.   Another proof, establishing the same formula by somewhat more strictly geometric methods, is given by Harvey.*   Unfortunately, the derivation of this result seems necessarily to be at best very roundabout, and we shall omit the details.

Enough has been said to indicate the diversity of the possible methods of proving this difficult theorem, and the lack of any simple proof.   In a later chapter we shall consider a series of more general theorems, bringing out some further and more significant aspects of the situation (**401** ff.).

325. We may note a few corollaries and extensions.

Theorem. *The four triangles of an orthocentric system determine sixteen inscribed and excribed circles.   These are all tangent to the common nine point circle of the system.*

All the tangent circles of the triangles discussed in **314** are tangent to the nine point circle.

Theorem. *A circle tangent to any three non-concurrent angle bisectors of the three angles of a triangle is tangent to the circumcircle of the triangle.*

We have determined the condition that a triangle can be inscribed in one given circle and circumscribed to another; and have seen that if one such triangle exists, there are an infinite number.   The centers of the nine point circles of such

---

* *Proceedings of Edinburgh Math. Society,* V, 1887, p. 102.

triangles are at a constant distance from I and therefore lie on a circle; and the nine point circles are all equal and therefore are tangent to two fixed circles.

The locus of the orthocenters of such triangles is a circle; the locus of the median points is a circle; and the three locus circles have their centers on OI, with O as a common homothetic center.

326. We may now study further the properties of Simson lines in a triangle, and establish simple relations associating them with the nine point circle. Further theorems about these lines will be developed at length in Chapter XIV.

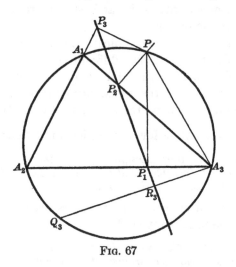

Fig. 67

Theorem. *The Simson line of any point on the circumcircle is perpendicular to the isogonals of the rays from the vertices to the given point.*

We know already that these isogonals are parallel **(234)**; but we have proved also **(237)** that for any point P of the plane, the isogonals of A_1P, A_2P, A_3P are respectively perpendicular to P_2P_3, P_3P_1, P_1P_2; and in the case before us, these are segments of one and the same line.

Corollary. *The Simson lines of diametrically opposite points of the circumcircle are mutually perpendicular. More generally, the angle between the Simson lines of two points equals the angle inscribed in the arc of the circumcircle between the points.*

Corollary. *The triangle formed by the Simson lines of three points is similar to the triangle of the points.*

Corollary. *There is a single point of the circumcircle whose Simson line is parallel to a given line. The construction for this point consists of drawing the perpendiculars from the vertices to the given line; the isogonals of these are concurrent at the desired point.*

327. Theorem. *The Simson line of any point of the circumcircle bisects the line joining the point to the orthocenter, and meets there the nine point circle.*

There seems to be no simple proof of this important theorem. Our proof (following Casey) consists in extending an altitude (fig. 68) A_1H to meet the circumcircle at H_1'; then if PH_1' cuts the base A_2A_3 at L_1 and the Simson line at X, we show that the Simson line is parallel to HL_1 and bisects PL_1 at X.

First, since P, P_1, P_2, A_3 are on a circle, we have

$$\angle P_2P_1P = \angle P_2A_3P = \angle A_1H_1'P = \angle P_1PH_1'$$

and triangle XPP_1 is isosceles, $\overline{PX} = \overline{XP_1}$. In other words, X is the mid-point of the hypotenuse of right triangle PP_1L_1. We have seen (**254**) that $\overline{HH_1} = \overline{H_1H_1'}$; hence

$$\angle HL_1H_1 = \angle H_1L_1H_1' = \angle P_1L_1P = \angle XP_1L_1$$

and HL_1 is parallel to P_1X. Therefore the Simson line P_1X bisects PH at S.

Finally, we have noted that the mid-point of every segment

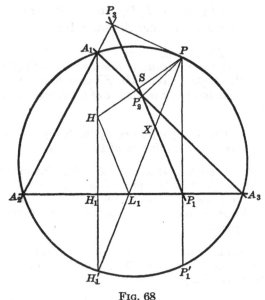

FIG. 68

HP from the orthocenter to the circumcircle is on the nine point circle; hence S is on the latter circle.

328. Theorem. *The Simson lines of two diametrically opposite points of the circumcircle intersect at right angles on the nine point circle.*

For if PQ is a diameter of the circumcircle, and if the midpoints of HP and HQ are S and T, then ST is, as we have just seen, a diameter of the nine point circle. But also the Simson lines of P and Q, as we have just seen, pass through S and T respectively, and are perpendicular to each other. Hence their intersection is on the nine point circle.

329. Theorem. *If the perpendicular from a point P of the circumcircle to any side A_2A_3 of the triangle is extended to meet the circumcircle at P_1', then A_1P_1' is parallel to the Simson line of P.*

For $\measuredangle PP_1P_2 = \measuredangle PA_3P_2 = \measuredangle PP_1'A_1$

which shows that P_1P_2 and $P_1'A_1$ are parallel.

330. We have proved (**268**) that the orthocenters of the triangles formed by four lines are collinear, and (**197**) that the circumcircles of the four triangles meet in a point whence the feet of the perpendiculars to the four lines are collinear.

Theorem. *The Simson line and the line of orthocenters of a complete quadrilateral are parallel, and the Simson line lies midway between the line of orthocenters and the common point of the circumcircles.*

For the line from this point to any orthocenter is bisected by the Simson line.

331. Theorem. *If four points are on a circle, there is a point through which pass the nine point circles of the four triangles, and the Simson line of each point with regard to the triangle of the other three.*

For if we join each point to the orthocenter of the triangle of the other three, the four connectors have a common midpoint (**265**). By **327**, each of the Simson lines and nine point circles passes through this point. This very suggestive theorem is generalized and extended in **400**.

332. Theorem. *Let four fixed points on a circle determine four triangles. Then any point of the circle determines four Simson lines, one for each triangle; the feet of the perpendiculars from the point to these four lines are collinear.*

For let $A_1A_2A_3A_4$ be any quadrangle inscribed in a circle, and P any other point on the circle. If P_{12} denotes the foot of the perpendicular from P to A_1A_2, etc., then the Simson line of P with regard to $A_1A_2A_3$ is $P_{23}P_{31}P_{12}$; and so on. We designate this line by l_4, and the foot of the perpendicular from P on it by T_4.

Now the triangle whose sides are l_1, l_2, l_3, and vertices are

P_{14}, P_{24}, P_{34} is inscribed in the circle drawn on A_4P as diameter. The Simson line of P with regard to this triangle passes through T_1, T_2, and T_3, which are therefore collinear. Obviously the fourth point T_4 must lie on the same line.

333. Theorem. *If a line through the orthocenter H cuts the sides of a triangle at L_1, L_2, L_3, then $H_1'L_1$, $H_2'L_2$, $H_3'L_3$, the reflections of the line with regard to the sides of the triangle, are concurrent at a point P of the circumcircle whose Simson line is parallel to the line $L_1L_2L_3$.*

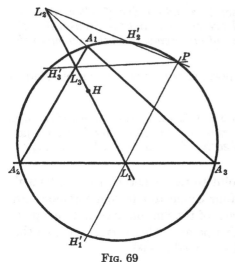

This is a by-product of the proof of **327**. Conversely:

Theorem. *If any point of the circumcircle be connected with the points H_1', H_2', H_3', the connectors meet the corresponding sides at three points collinear with H; this line is parallel to the Simson line of the given point.*

FIG. 69

334. Theorem. *If PQ is a diameter of the circumcircle, and if the perpendiculars from P and Q to their respective Simson lines meet at R, then R is on the circumcircle and its Simson line is parallel to PQ.*

335. Theorem. *If from any vertex of a triangle, perpendiculars are drawn on the interior and exterior bisectors of the other angles, the feet of these perpendiculars are on a line, which bisects the adjacent sides of the triangle.*

336. Theorem. *Of the arcs intercepted on the nine point circle between the Simson lines of two points, one is double the other.*

That is, if A, B are on the circumcircle, and A', B' are the mid-points of AH and BH respectively; and if the Simson lines of A and B cut the nine point circle at A', C and at B', D respectively; then one of the arcs CD is double one of the arcs $A'B'$.

For arc $A'B'$ is similar to arc AB; and also the angle between the Simson lines, which is equal to the angle inscribed in this arc, is measured by the sum or the difference of $A'B'$ and CD.

337. Numerous questions concerning Simson lines suggest themselves. For example, how many Simson lines are there through a given point? Under what conditions are the Simson lines of three points concurrent? We shall briefly summarize some of the further properties of these interesting lines.*

Theorem. *Through the intersection of any two Simson lines there passes a third; thus the points of the circumcircle are associated in threes, any two points determining a third partner.*

Let P and Q be any points of the circumcircle; since their Simson lines are not parallel, let them intersect at S. Extending HS by its own length to H', and denoting by R the orthocenter of PQH', we prove that R lies on the given circle $A_1A_2A_3PQ$. (For $\angle PRQ = \angle QH'P$, which equals the angle between the Simson lines of P and Q, since these are parallel respectively to $H'P$ and $H'Q$; and therefore equals $\angle PA_1Q$.) Similarly we prove that the Simson line of R is parallel to RH', and passes through S. By such methods, we can obtain without difficulty all the following results (see further **406**).

338. Theorem. *Given a triangle $A_1A_2A_3$ and two points P, Q of its circumcircle; there exists a third point R of the circle, such that the Simson lines of P, Q, and R are concurrent. The point of concurrence is midway between the orthocenters of*

* See, for instance, Beard, *Educational Times Reprint*, II, 20, p. 109.

triangles $A_1A_2A_3$ and PQR, and the Simson lines of A_1, A_2, A_3 with respect to triangle PQR are concurrent at the same point S. To locate R, HS is extended by its own length to H'; then R is the orthocenter of triangle PQH'. The Simson line of each of the points P, Q, R is perpendicular to the line joining the other two. Conversely, if P is any point of the circumcircle, and QR a chord perpendicular to the Simson line of P, then the Simson lines of P, Q, and R are concurrent.

339. Problem. *To determine the Simson line or lines of a given triangle, which pass through a given point.*

We have seen (**326**) how to solve this problem when the given point is at infinity. In general, however, it cannot be solved by elementary methods. Analytically it can be shown that all the Simson lines are tangent to a certain curve with three equidistant cusps (hypocycloid) which is circumscribed about the nine point circle. Through a point outside this curve there is one Simson line, and through an interior point there are three.

Exercise. Draw the Simson lines of a large number of points, and observe this hypocycloid curve.

Exercise. (Sanjana, *Educational Times Reprint*, II, p. 3.) If P, Q, and R are on the circumcircle, with QR parallel to the Simson line of P, then PR and PQ are parallel respectively to the Simson lines of Q and R. The triangle whose sides are the Simson lines is similar to PQR and similarly placed; determine the locus of the center of similitude as QR moves parallel to the fixed Simson line of P.

Exercise. Complete all indicated proofs in this chapter, namely: **310–313, 316–319, 325, 326, 331, 333, 334, 335, 338, 339.**

CHAPTER XII

THE SYMMEDIAN POINT AND OTHER NOTABLE POINTS

340. In the foregoing chapters we have discussed the properties of the best known of the notable points, lines, and circles of the triangle. Another whole system, associated with the name of Brocard, is deserving of detailed study, and will be taken up in Chapters XVI, XVII, and XVIII. This configuration is to some extent independent of that which we have studied; but a liaison is furnished by the symmedian point, which we now propose to study, and which bears important relations both to the configuration of orthocenter, circumcenter, and median point, and to that which is based on the Brocard points. After discussing the most interesting properties of this point, we devote the latter part of the chapter to some other notable points of secondary importance.

341. Definition. The line isogonal to any median of a triangle is called a *symmedian*. The point of concurrence of the symmedians, the isogonal conjugate of the median point, is called the *symmedian point K*.

This point has various names; by French and British writers it is usually called Lemoine's point, and by German, Grebe's. While both Lemoine and Grebe have contributed to our knowledge of the point, neither was the first discoverer, and the neutral descriptive term seems preferable. According to a thorough study by Mackay,* the point was not discovered at any one time, but gradually came into prominence by

* "Early History of the Symmedian Point," *Proceedings of Edinburgh Math. Society*, XI, 1892–93, p. 92.

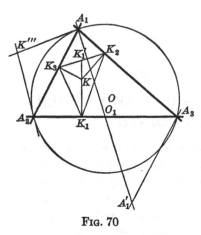

FIG. 70

the recognition of its various properties by different investigators.

342. Theorem. *The perpendiculars on the sides of the triangle from the symmedian point are proportional to the lengths of the sides (235, 272). Conversely, the only point within a triangle from which the lengths of the perpendiculars to the sides are proportional to the lengths of the sides is the symmedian point.*

Corollary. $\overline{KK}_1 = a_1 \dfrac{2\,\Delta}{a_1^2 + a_2^2 + a_3^2}$, etc.

For we may write $\overline{KK}_1 = c\,a_1$, $\overline{KK}_2 = c\,a_2$, $\overline{KK}_3 = c\,a_3$ and determine c from the equation

$$2\,\Delta = a_1 \cdot \overline{KK}_1 + a_2 \cdot \overline{KK}_2 + a_3 \cdot \overline{KK}_3 = c\,(a_1^2 + a_2^2 + a_3^2)$$

343. Theorem. *A line through a vertex of the triangle and tangent to the circumcircle is isogonal to the line parallel to the opposite side of the triangle (exmedian). From a point on the tangent, the perpendiculars to the sides of the angle are proportional to the lengths of those sides.*

344. Definition. The tangents at the vertices of the triangle to the circumcircle are called *exsymmedians* (**277**).

Theorem. *Any two exsymmedians and the third symmedian are concurrent at a point, an exsymmedian point (cf.* **224***). The perpendiculars from an exsymmedian point K' to the sides are proportional to the lengths of the sides (***235***);*

$$\overline{K'K_1}' = a_1 \dfrac{2\,\Delta}{a_2^2 + a_3^2 - a_1^2}, \text{ etc.}$$

Theorem. *A symmedian divides the opposite side internally, and an exsymmedian divides it externally, in the ratio of the squares of the adjacent sides* (**244** *a* or **84**).

Obviously any theorem concerning the symmedian point can be paralleled by a theorem concerning each exsymmedian point.*

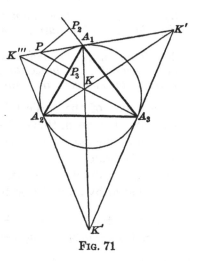

345. Theorem. *A line parallel to a symmedian, and terminated by the adjacent sides, is bisected by the corresponding exsymmedian.*

346. Theorem. *A line antiparallel to a side of a triangle is bisected by the symmedian from the corresponding vertex.*

Fig. 71

For if we turn the figure over on itself, the symmedian becomes a median, and the antiparallel becomes parallel to the opposite side.

Corollary. *The lines joining the vertices of a triangle to the mid-points of the corresponding connectors of the feet of the altitudes are concurrent at the symmedian point.*

This yields a convenient practical construction for K. Another, based on **344**, follows; still others are given in **451** and **452**.

347. Construction. *If the tangents to the circumcircle at A_1, A_2, A_3 meet at K', K'', K''', then A_1K', A_2K'', A_3K''' are concurrent at the symmedian point K.*

---

* An extensive treatment of the symmedians is that of Mackay, "Symmedians of a Triangle and their Concomitant Circles," *Proceedings of Edinburgh Math. Society*, XIV, 1896, pp. 37–103.

Theorem. *The Gergonne point of a triangle (291 a) is the symmedian point of the triangle whose vertices are the points of contact of the inscribed circle.*

348. Theorem. *The symmedian point of a right triangle is the mid-point of the altitude on the hypotenuse.*

349. Theorem. *The point in the plane, for which the sum of the squares of the distances to the sides of a triangle is a minimum, is the symmedian point K.*

This is one of the earliest known properties of the symmedian point; indeed, this sort of thing occupied much of the attention of the early geometers (cf. **264, 275**). The easiest proof is algebraic.

For any six quantities whatever, the following identity can be established by actually performing the multiplications.

$$(a_1^2 + a_2^2 + a_3^2)(x_1^2 + x_2^2 + x_3^2) = (a_1x_1 + a_2x_2 + a_3x_3)^2 \\ + (a_2x_3 - a_3x_2)^2 + (a_3x_1 - a_1x_3)^2 + (a_1x_2 - a_2x_1)^2$$

Now let a_1, a_2, a_3 designate the sides of the triangle, x_1, x_2, x_3 the signed perpendiculars from any point; then $(a_1x_1 + a_2x_2 + a_3x_3)$ represents twice the area of the triangle, and is constant. Since every term is positive or zero, we shall have $x_1^2 + x_2^2 + x_3^2$ a minimum when the last three terms are zero, that is, when

$$x_1 : x_2 : x_3 = a_1 : a_2 : a_3$$

The required point is therefore either the symmedian point or an exsymmedian point.

By comparing the formulas for the respective perpendiculars (**342, 344**), we see that the symmedian point furnishes the required minimum, viz:

$$\overline{KK_1}^2 + \overline{KK_2}^2 + \overline{KK_3}^2 = \frac{4\,\Delta^2}{a_1^2 + a_2^2 + a_3^2}$$

350. Theorem. *The symmedian point is the median point of its own pedal triangle.*

For if the median A_1O_1 is extended by its own length to A_1' (figure 70), triangles KK_2K_3 and $A_3A_1A_1'$ have corresponding sides perpendicular each to each (**237**) and are therefore similar. If KK_1 meets K_2K_3 at K_1', then KK_1' is homologous to A_3O_1 in these similar figures, and K_1' is the midpoint of K_2K_3. Thus K_1K_1' is a median of triangle $K_1K_2K_3$, and K is its median point.

Theorem. *Conversely, if through the vertices of a triangle lines are drawn perpendicular to the medians, the median point of the given triangle is the symmedian point of the new triangle.*

Corollary. *The symmedian point is the only point which is the median point of its own pedal triangle.*

Theorem. *Similarly, an exsymmedian point is an exmedian point of its own pedal triangle.*

That is, if K' is an intersection of two of the tangents to the circumcircle, K' is one vertex of a parallelogram whose other vertices are also those of its pedal triangle.

351. Theorem. *Of all triangles inscribed in a given triangle, that for which the sum of the squares of the sides is a minimum is the pedal triangle of the symmedian point.*

For let X_1, X_2, X_3 be any points on A_2A_3, A_3A_1, A_1A_2 respectively; let P be the median point of $X_1X_2X_3$, and let $P_1P_2P_3$ be the pedal triangle of P. If perchance P_1, P_2, P_3 coincide respectively with X_1, X_2, X_3, then P is the symmedian point K. In every other case,

$$\overline{PP_1}^2 + \overline{PP_2}^2 + \overline{PP_3}^2 < \overline{PX_1}^2 + \overline{PX_2}^2 + \overline{PX_3}^2$$

But we saw in **349** that if K is the symmedian point,

$$\overline{KK_1}^2 + \overline{KK_2}^2 + \overline{KK_3}^2 < \overline{PP_1}^2 + \overline{PP_2}^2 + \overline{PP_3}^2$$

Now in each of the triangles $K_1K_2K_3$ and $X_1X_2X_3$, we may apply **96** c, whence the result.

Theorem. *If the symmedians meet the circumcircle at L_1, L_2, L_3, then K is also the symmedian point of triangle $L_1L_2L_3$.* (**199, 244** c, **350**)

The triangles $A_1A_2A_3$ and $L_1L_2L_3$, having the same circumcircle, symmedians, and symmedian point, are called co-symmedian triangles. We shall later consider them again (**475**).

Exercise. *If squares are constructed externally on the sides of a triangle, the sides parallel to those of the given triangle form a triangle similar to the latter, with K as center of similitude.*

Problem. *Given a vertex of a triangle, the directions of the sides through it, and the symmedian point, to construct the triangle.*

THE ISOGONIC CENTERS *

352. The next points to occupy our attention are the *isogonic centers*, which have the property that from either of them the angles subtended by the sides are 60° or 120°. They are determined by the following theorem.

Theorem. *If equilateral triangles $A_2A_3P_1$, $A_3A_1P_2$, $A_1A_2P_3$ are constructed externally to the given triangle, then $\overline{A_1P_1}$, $\overline{A_2P_2}$, $\overline{A_3P_3}$ are equal; they are concurrent at a point R, and*

$$\angle A_2RA_3 = \angle A_3RA_1 = \angle A_1RA_2 = 120°$$

For triangles $A_1A_2P_1$ and $P_3A_2A_3$ are congruent; A_2 is their center of similitude, and the angle between corresponding lines is equal to $P_1A_2A_3$, or 60°. Thus A_1P_1 and A_3P_3 are equal, and if they meet at R, then

$$\angle A_3RA_1 = \angle A_3P_3, A_1P_1 = \angle A_2A_3, A_2P_1 = \angle A_2P_1A_3$$

* The rest of this chapter may be omitted without embarrassment, with the exception of 356–358, which will be used later.

whence it follows that R is the intersection of the circles $A_2A_3P_1$, $A_1A_2P_3$. Since the circle $A_1A_3P_2$ passes through this point also, therefore A_2P_2 passes through R, as was to be

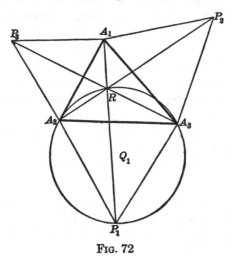

Fig. 72

proved.   We see that $\measuredangle A_2RA_3$ is 120°, rather than 60°, because triangles $A_1A_2A_3$ and $P_1A_2A_3$ are described in opposite directions, and

$$\measuredangle A_2RA_3 = \measuredangle A_2P_1A_3 = 120°$$

The point R will fall within the triangle, and the angles subtended by the respective sides will actually be 120°, except when one of the given angles exceeds 120°.   In this case R is outside the triangle, and the shorter sides subtend angles of 60°.

Similarly:

Theorem. *If equilateral triangles $A_2A_3P_1'$, $A_3A_1P_2'$, $A_1A_2P_3'$ are constructed internally on the sides of a non-equilateral triangle, then A_1P_1', A_2P_2', A_3P_3' are equal, and meet at a point R', whence the sides subtend equal angles,*

$$\angle A_2R'A_3 = \angle A_3R'A_1 = \angle A_1R'A_2 = 60°$$

The proof resembles that of the preceding theorem. For a triangle having no angle of 60° and only one angle greater than 60°, R' lies outside the triangle, opposite this angle. If, however, there are two angles greater than 60°, R' is opposite to the third angle, and outside the triangle. (For an equilateral triangle there is no point R'; but if a scalene triangle changes its form so as to become equilateral, R' approaches a limiting position which may be anywhere on the circumcircle. Let the reader investigate.)

The points R and R' are the only ones at which each side of the triangle subtends angles of 120° or 60°. There are other points from which two sides subtend such angles; what are they?

353. Theorem. *The algebraic sum of the distances from an isogonic center to the vertices of the triangle equals the length of the equal segments from the latter to the vertices of the equilateral triangles:*

a. *In case R is inside the triangle,* $\overline{A_1P_1} = \overline{A_1R} + \overline{A_2R} + \overline{A_3R}$

b. *If $A_3 > 120°$,* $\overline{A_1P_1} = \overline{A_1R} + \overline{A_2R} - \overline{A_3R}$

c. *If R' is opposite to A_3,* $\overline{A_1P_1'} = \overline{A_3R'} - \overline{A_1R'} - \overline{A_2R'}$

For R is on the circle $P_1A_2A_3$, hence by **93 b**

$$\overline{P_1R} = \overline{A_2R} + \overline{A_3R}$$

354. *a.* **Theorem.** *Triangles $A_1P_3P_2'$, $A_1P_3'P_2$, $A_1A_2A_3$ are congruent. Corresponding lines meet at angles of 60°.*

b. **Theorem.** *A_1P_2 and A_2P_1' meet on the circumcircle.*

c.
$$\overline{A_1P_1}^2 + \overline{A_1P_1'}^2 = a_1^2 + a_2^2 + a_3^2 \tag{96}$$
$$\overline{A_1P_1}^2 - \overline{A_1P_1'}^2 = 4\sqrt{3}\,\Delta \tag{98}$$

d.
$$\overline{A_1P_1}^2 = \tfrac{1}{2}(a_1^2 + a_2^2 + a_3^2) + 2\sqrt{3}\,\Delta$$
$$\overline{A_1P_1'}^2 = \tfrac{1}{2}(a_1^2 + a_2^2 + a_3^2) - 2\sqrt{3}\,\Delta$$

(Nicholas Fuss, 1796)

e. Further study of the circles circumscribed about the equilateral triangles $A_2A_3P_1$, $A_2A_3P_1'$, etc., is suggested. *If the centers of these circles are Q_1, Q_1', etc., then triangles $Q_1Q_2Q_3$ and $Q_1'Q_2'Q_3'$ are equilateral, with the median point M as common center. The isogonic centers R' and R are respectively on the circles $Q_1Q_2Q_3$ and $Q_1'Q_2'Q_3'$.*

355. We consider next a property of the isogonic center R which lends it a singular historic interest. This point was indeed the first notable point of the triangle discovered in more recent times than that of Greek mathematics. In the seventeenth century, Fermat proposed to Torricelli the problem of determining a point whose distances from three fixed points should have the minimum sum. The practical application of this problem is obvious. Torricelli solved the problem, and thus discovered the point R. His solution was published in 1659 by his pupil Viviani. The elegantly simple analysis of the problem which follows is due to Steiner.

Theorem. *If $A_1A_2A_3$ is a triangle having no angle as great as $120°$, the point the sum of whose distances from the vertices of the triangle is the isogonic center R.*

If we connect either isogonic center with the vertices, and draw lines through the latter perpendicular to the connectors, we form an equilateral triangle $X_1X_2X_3$. It is clear that the isogonic center will be inside the triangle $X_1X_2X_3$, only if it is the point R and is within the given triangle.

Now the algebraic sum of the perpendiculars from a moving point to the sides of a fixed equilateral triangle is constant; if the point is outside the triangle, the numerical sum is greater than the algebraic sum. Now let S be any point in the plane, except R, and denote by s_1, s_2, s_3 the perpendiculars on X_2X_3, X_3X_1, X_1X_2 from S. Then

$$s_1 + s_2 + s_3 \geqq \overline{RA_1} + \overline{RA_2} + \overline{RA_3}$$

according as S is inside or outside triangle $X_1X_2X_3$.

In any case, $\overline{SA} \geqq s_1,\ \overline{SA}_2 \geqq s_2,\ \overline{SA}_3 \geqq s_3$,
and therefore

$$\overline{SA}_1 + \overline{SA}_2 + \overline{SA}_3 > \overline{RA}_1 + \overline{RA}_2 + \overline{RA}_3$$

Similarly, *the triangle $X_1X_2X_3$ is the largest equilateral triangle that can be described in the positive sense with its sides passing through the vertices of the given triangle.*

If any angle of a triangle exceeds 120°, the vertex of this angle is the solution of Fermat's problem.

The isogonal conjugates of the isogonic centers are called the *isodynamic points,* and will be taken up in detail in chapter XVII.*

356. The theorems concerning the isogonic centers are susceptible of some generalization. We shall state without proof a number of theorems, beginning with the most general and indicating several forms of modification. The proofs depend on the theorem of Ceva.

Theorem.† *From each vertex of a triangle let a pair of isogonal rays be drawn, each associated with one of the adjacent sides of the angle. Let the rays associated with each side meet at a point, which is then connected with the opposite vertex. The three connectors are concurrent. That is, if $\measuredangle A_3A_1X_2 = \measuredangle X_3A_1A_2 = \phi_1,\ \measuredangle A_1A_2X_3 = \measuredangle X_1A_2A_3 = \phi_2,\ \measuredangle A_2A_3X_1 = \measuredangle X_2A_3A_1 = \phi_3$, then A_1X_1, A_2X_2, A_3X_3 meet at a point P, from which the perpendiculars to the sides of the triangle are given by*

$$p_1 : p_2 : p_3 = \frac{\sin \phi_1}{\sin(\alpha_1 - \phi_1)} : \frac{\sin \phi_2}{\sin(\alpha_2 - \phi_2)} : \frac{\sin \phi_3}{\sin(\alpha_3 - \phi_3)}$$

In the particular case that

$$\phi_1 + \phi_2 + \phi_3 = 180°$$

the triangles constructed on the sides are similar, in the sense

$$A_1A_2X_3 \sim X_1A_2A_3 \sim A_1X_2A_3$$

---

* A detailed treatment of the isogonic centers is given by Mackay, *Proceedings of Edinburgh Math. Society,* XV, 1897, pp. 100–18.

† Fig. 73 illustrates this theorem for the special case of **357.**

If A_1X_1, A_2X_2, A_3X_3 meet at P, the circumcircles of the three triangles also pass through P; the angles formed at P are equal to the given angles ϕ_1, ϕ_2, ϕ_3; the lengths $\overline{A_1X_1}$, $\overline{A_2X_2}$, $\overline{A_3X_3}$ are proportional to the altitudes of these triangles.

357. Specializing the original theorem in another manner we are led to the following. The theorems relating to the isogonic centers are cases of these theorems; other cases will present themselves in the Brocard geometry.

Theorem. *If similar isosceles triangles are drawn in similar position, with the sides of the given triangle as bases, the lines connecting their vertices with the opposite vertices of the triangle are concurrent. If ϕ is the given base angle, the distances from the point of concurrence to the sides are given by*

$$p_1 : p_2 : p_3 = \frac{1}{\sin(\alpha_1 - \phi)} : \frac{1}{\sin(\alpha_2 - \phi)} : \frac{1}{\sin(\alpha_3 - \phi)}$$

Conversely, if the perpendiculars from a point are proportional to such expressions as these, the point determines such a set of similar isosceles triangles.

This theorem is an immediate consequence of the theorem of Ceva. Special cases are: $\phi = 0$, the median point; $\phi = 90°$, the orthocenter; $\phi = \alpha_1$, the vertex A_1; $\phi = 60°$ or $120°$, the isogonic centers; and so on. It is instructive to sketch the figure for varying values of ϕ, and determine the locus of the point of concurrence.

358. Theorem. *If triangles $A_2A_3X_1$, $A_3A_1X_2$, $A_1A_2X_3$ are directly similar and similarly placed, the median point of $X_1X_2X_3$ coincides with that of $A_1A_2A_3$.*

Let X_1O_1 be extended by its own length to X_1', so that $A_2X_1A_3X_1'$ is a parallelogram. Then triangles $A_1A_2A_3$, $X_2X_1'A_3$, $X_3A_2X_1'$ are similar, having in each case a pair of equal angles and the including sides proportional. Then by proportions we prove that $\overline{A_1X_2} = \overline{X_3X_1'}$, $\overline{A_1X_3} = \overline{X_2X_1'}$, and $A_1X_3X_1'X_2$ is a parallelogram. Thus A_1X_1' and X_2X_3

---

* Note that this theorem does not assume that the similar triangles are isosceles nor that A_1X_1, A_2X_2, A_3X_3 are concurrent.

bisect each other, say at Z. Next, two medians A_1O_1 and X_1Z of triangle $A_1X_1X_1'$ trisect each other at M, since A_1O_1 is also a median of $A_1A_2A_3$; but X_1Z is also a median of $X_1X_2X_3$, and therefore the median point of the latter is also at M.

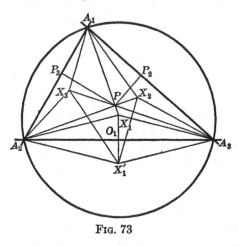

FIG. 73

A trigonometric proof may be built up by expressing the perpendiculars from X_1, X_2, X_3 to some one side, say A_2A_3, in terms of the angles of the triangle and the angles $X_1A_2A_3$ and $X_1A_3A_2$; by manipulation of trigonometric expressions it is easily proved that the sum of these perpendiculars is equal to A_1H_1, so that their average equals MM_1.

Corollary. *If $X_1X_2X_3$ are collinear, their line passes through the median point.*

359. Theorem. *From each vertex of a triangle let a pair of isogonal rays be drawn. If no three of these are concurrent, they intersect at twelve points besides the vertices. These twelve points are isogonally conjugate in pairs; each vertex of the given triangle may be connected with two pairs of them by new lines. These new lines intersect by threes in eight other points, which are four pairs of isogonal conjugates.*

This figure will amply repay extensive study. For instance, we find that the lines connecting the six pairs of isogonal conjugates meet by threes at four points, and those connecting the last named four pairs of isogonal conjugates are concurrent.

360. Antipedal triangles. The method used in the proof of **355** suggests an interesting generalization.

Definition. *If any point is connected with the vertices of a triangle, the lines through the vertices and perpendicular to these connectors are the sides of a triangle called the antipedal triangle of the point with regard to the given triangle.*

It is evident that the given triangle is the pedal of the point with regard to this antipedal triangle, whence the name. Thus any theorem concerning pedal triangles suggests a theorem on antipedal triangles.*

Theorem. *The antipedal triangle of a point on the circumcircle reduces to a point, also on the circumcircle.*

Theorem. *The antipedal triangle of a point is homothetic to the pedal triangle of its isogonal conjugate.*

THE NAGEL POINT

361. We have already alluded (**291**) to the point of Nagel, defined as the intersection of the lines from the vertices of the triangle to the points of contact of the opposite escribed circles. We now establish some of its interesting properties.

Theorem. *The median point M, the incenter I, and the Nagel point N are collinear, and $\overline{MN} = 2\,\overline{IM}$.* (Cf. **257**.)

We first consider right triangles $A_1 H_1 J_1'$ and $I I_1 O_1$.

$$\overline{I_1 I} = \rho = \frac{\Delta}{s}, \quad \overline{A_1 H_1} = \frac{2\,\Delta}{a_1}, \quad \overline{I_1 O_1} = \tfrac{1}{2}\,(a_2 - a_3)$$

$$\overline{J_1' H_1} = s - a_3 - \overline{A_2 H_1} = \frac{s\,(a_2 - a_3)}{a_1}$$

Therefore
$$\frac{\overline{J_1' H_1}}{\overline{O_1 I_1}} = \frac{\overline{A_1 H_1}}{\overline{I_1 I}} = \frac{2\,s}{a_1}$$

* See further Gallatly, *l.c.*, Chap. VII.

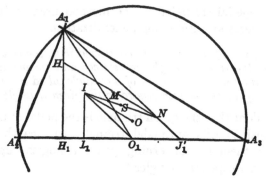

so that the right triangles are similar, and IO_1 is parallel to A_1N. By application of **84** and **290** we find that

$$\frac{\overline{A_1J_1'}}{\overline{A_1N}} = \frac{s}{a_1}$$

so that $\overline{A_1N}$ is $2\,\overline{IO_1}$. Hence A_1O_1 and IN trisect each other at M.

362. Theorems. *HN is parallel to OI, $\overline{HN} = 2\,\overline{IO}$.*
If S is the mid-point of IN, then O_1S is parallel to A_1I, and therefore is a bisector of the angle $O_1O_2O_3$. Thus S is the in-center of triangle $O_1O_2O_3$.
The centers of similitude of the circles inscribed in triangles $A_1A_2A_3$ and $O_1O_2O_3$ are M and N.

363. Theorem. *The lines from the vertices of the triangle to the Nagel point pass through the respective points of tangency of the incircle of triangle $O_1O_2O_3$.*

Theorem. *The incenter is the Nagel point of $O_1O_2O_3$.*

364. The circle inscribed in triangle $O_1O_2O_3$ is called the *P-circle* or *Spieker circle,** and has some properties curiously parallel to those of the nine point circle. We have just located its center S as the mid-point on IN, and seen that the four

---

* Spieker, *Grunerts Archiv*, 51, 1870, pp. 10–14.

points N, S, M and I are placed similarly to H, F, M and O. Also the lines from A_1, A_2, A_3 to N pass through the respective points of contact of the Spieker circle

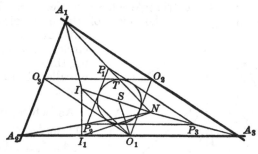

FIG. 75

365. Theorem. *The Spieker circle is also inscribed in the triangle whose vertices are midway from the Nagel point to the vertices of the triangle. Specifically, if $P_1P_2P_3$ are the mid-points of A_1N, A_2N, A_3N, then P_2P_3, IO_1 NI_1, meet at a point where the Spieker circle is tangent to P_2P_3.*

Obviously triangles $A_1A_2A_3$ and $P_1P_2P_3$ are similar, in the ratio $2:1$, with N as homothetic center; hence the incircle of the second is precisely the Spieker circle. Also the lines NI_1, etc., pass through the points of contact of the latter, and are there bisected. Finally, IO_1 is parallel to A_1NJ_1', and bisects I_1J_1' at O_1; hence it passes through the mid-point of I_1N.

366. Theorem. *The Spieker circle is inscribed in two congruent triangles $O_1O_2O_3$ and $P_1P_2P_3$; the points of contact of corresponding sides of these triangles are diametrically opposite points.*

Thus we see that whereas the nine point circle is half the size of the circumcircle, with M as a center of similitude, and has six notable diameters, so the Spieker circle is half the size of the incircle, with M as a center of similitude, is tangent to six notable lines at notable points, and thus has six notable diameters.

367. Definition. The *Fuhrmann Triangle (Spiegeldreieck*)* of a triangle is the triangle whose vertices are the reflections, with regard to the sides, of the mid-points of the outer arcs of the circumcircle. The circle passing through these points is called the *Fuhrmann circle.*

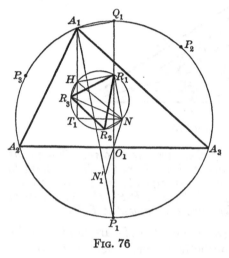

As before, we denote the mid-points of arcs A_2A_3, A_3A_1, A_1A_2 by P_1, P_2, P_3, so that A_1, I, P_1 are collinear; and the diametrically opposite points by Q_1, Q_2, Q_3. If the reflections of P_1, P_2, P_3 with regard to the respective sides are R_1, R_2, R_3,

FIG. 76

then $$\overline{R_1O_1} = \overline{O_1P_1} = \tfrac{1}{2} a_1 \tan \frac{\alpha_1}{2}.$$

368. Theorem. $A_1HR_1Q_1$ *is a parallelogram, and HR_1 is perpendicular to the angle bisector A_1IP_1.*

For $\overline{Q_1R_1} = 2\,\overline{OO_1} = \overline{A_1H}$ (**256**); and therefore HR_1 is parallel to A_1Q_1, which bisects the external angle.

369. Theorem. NH *is a diameter of the Fuhrmann circle.*

First we extend NO_1 to meet the bisector A_1IP_1 at N_1'. By virtue of the proof of **361**, O_1 is the mid-point of NN_1', and $NP_1N_1'R_1$ is a parallelogram. Therefore R_1N is parallel

* Fuhrmann, *l.c.*, p. 107.

to A_1I, whereas we have seen that HR_1 is perpendicular to it. We thus have R_1 on the circle whose diameter is HN.

370. Theorem. *Triangles $P_1P_2P_3$ and $R_1R_2R_3$ are inversely similar.*

For R_1H is parallel to P_2P_3, etc., and we have

$$\angle R_1R_2R_3 = \angle R_1HR_3 = \angle P_2P_3,\ P_1P_2 = \angle P_3P_2P_1$$

371. Theorem. *The Fuhrmann circle cuts each altitude at a distance 2ρ from the vertex.*

For on the altitude A_1H_1, let us cut off A_1T_1 equal to 2ρ; then since

$$2\Delta = a_1 \cdot \overline{A_1H_1} = 2\rho s$$

we have $\qquad \dfrac{\overline{A_1T_1}}{\overline{A_1H_1}} = \dfrac{a_1}{s} = \dfrac{\overline{A_1N}}{\overline{A_1J_1}},$ $\qquad$ (361)

showing that T_1N is parallel to A_2A_3, and HT_1N is a right angle.

372. Theorem. *Triangles $A_1A_2A_3$ and $T_1T_2T_3$ are inversely similar.*

We thus have eight notable points on this interesting circle. Many other remarkable properties are worked out by the discoverer.

In conclusion, we may note that while our attention has been confined to the figures associated with the incircle, there exist corresponding configurations bearing the same relations to the excircles. That is, there exist three "ex-Nagel-points" having properties like those of the Nagel point, and each giving rise to a Spieker circle and a Fuhrmann circle. The more detailed investigation of these figures will be found well worth while.

Exercise. Of the following theorems in this chapter, left to be proved by the student in whole or in part, some will be found more difficult than those of earlier chapters; **342–348, 351, 353, 354, 356, 357, 359, 360, 362, 363, 366.**

CHAPTER XIII

TRIANGLES IN PERSPECTIVE

373. In this chapter * we consider briefly a number of theorems of a projective nature; that is, theorems relating to the concurrence of lines and the collinearity of points, but not to distances, angles, or ratios. We discuss the relationship of perspective figures, and establish the fundamental theorem of Desargues. Various applications of this theorem, followed by a brief study of the quadrilateral, lead us finally to the famous theorem of Pascal concerning the hexagon inscribed in a circle.

Definition. Two figures in a plane are said to be *in perspective* with each other, when and if: (*a*) the lines joining corresponding points are concurrent at a point called the *center of perspective;* and (*b*) the points of intersection of corresponding lines are on a line called the *axis of perspective*.

We have already considered a special case under this definition, namely that of similar figures whose corresponding sides are parallel. In this case the axis of perspective is the line at infinity, and the center of perspective is the homothetic center. The existence of figures in perspective in the more general case is established by the following theorem.

374. Theorem of Desargues. *If two triangles have a center of perspective, they have an axis of perspective.*

Let triangles $A_1A_2A_3$ and $B_1B_2B_3$ be so placed that A_1B_1, A_2B_2, A_3B_3 meet at a point O; and let A_2A_3 and B_2B_3 meet at

---

* Any parts or all of Chapters XIII, XIV, XV may be omitted without impairing the sequence of later chapters. The present chapter, however, contains several famous theorems.

C_1, A_3A_1 and B_3B_1 at C_2, A_1A_2 and B_1B_2 at C_3. The proof that C_1, C_2 and C_3 are collinear is easily effected by the

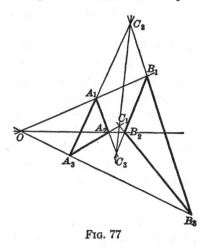

FIG. 77

theorem of Menelaus. First we consider the line $B_3C_2B_1$ as a transversal of triangle A_1OA_3

$$\frac{\overline{B_3O} \cdot \overline{C_2A_3} \cdot \overline{B_1A_1}}{\overline{B_3A_3} \cdot \overline{C_2A_1} \cdot \overline{B_1O}} = 1$$

Similarly

$$\frac{\overline{B_1O} \cdot \overline{C_3A_1} \cdot \overline{B_2A_2}}{\overline{B_1A_1} \cdot \overline{C_3A_2} \cdot \overline{B_2O}} = 1$$

and

$$\frac{\overline{B_2O} \cdot \overline{C_1A_2} \cdot \overline{B_3A_3}}{\overline{B_2A_2} \cdot \overline{C_1A_3} \cdot \overline{B_3O}} = 1$$

Combining these equations and canceling, we have

$$\frac{\overline{C_1A_2} \cdot \overline{C_2A_3} \cdot \overline{C_3A_1}}{\overline{C_1A_3} \cdot \overline{C_2A_1} \cdot \overline{C_3A_2}} = 1$$

which shows that C_1, C_2, C_3 are collinear.

375. Theorem. *Conversely, if two triangles have an axis of perspective, they have a center of perspective.*

With the same notation, we assume that C_1, C_2, C_3 are on a line, and show that A_1B_1, A_2B_2, A_3B_3 are concurrent. Consider triangles $A_1B_1C_3$ and $A_3B_3C_1$; they have C_2 as center of perspective, and therefore by the preceding theorem they have an axis of perspective. In other words, the line through A_2 and B_2 passes also through the point of intersection of A_1B_1 and A_3B_3, which is just what we wished to prove.

Evidently there are various special cases, when some of the lines are parallel. The proofs are easily modified for these cases, and hereafter we shall not distinguish them.

We see then that two triangles will be in perspective when they have either a center or an axis of perspective. For figures other than triangles, we may easily establish the following general statement.

376. Theorem. *Given a center of perspective O, an axis of perspective p, a figure ABC . . ., and a point A' on OA, there exists, and can be constructed with ruler only, a figure in perspective with the given figure, in which A' corresponds to A.*

To find the corresponding point B' to any point B, let AB meet p at M, and draw MA', meeting OB at B'. Then by Desargues' theorem, the point corresponding to any other point C is definitely and uniquely determined.

377. Theorem. *If three triangles have a common center of perspective, their three axes of perspective are concurrent.*

Let the triangles be $X_1X_2X_3$, $Y_1Y_2Y_3$, $Z_1Z_2Z_3$, so that $X_1Y_1Z_1$, $X_2Y_2Z_2$, and $X_3Y_3Z_3$ are concurrent straight lines. We shall denote any side of a triangle by a small letter, the same as that of the opposite vertex. Then consider the triangles whose sides are x_2, y_2, z_2 and x_3, y_3, z_3. Corresponding sides meet at the collinear points X_1, Y_1, Z_1; hence lines joining corresponding vertices are concurrent. But the line joining the intersection of x_2 and y_2 with that of x_3 and y_3 is the axis of perspective of $X_1X_2X_3$ and $Y_1Y_2Y_3$; and so on. Thus the three axes are concurrent.

378. Theorem. *Similarly, if three triangles are in perspective two by two, and on a common axis of perspective, their centers of perspective are collinear.*

The proofs of this and the various converse theorems present no difficulty.

379. The application of the theorem of Menelaus to the points of intersection of the sides of two triangles yields a general formula which can be applied in a number of ways to special cases. Given two triangles, $A_1A_2A_3$ and $B_1B_2B_3$; let A_2A_3 be cut by B_2B_3 at P_1, by B_3B_1 at Q_1, and by B_1B_2 at R_1; and let the intersections on A_3A_1 and A_1A_2 be similarly designated. Then on the line B_2B_3 are the intersections P_1, R_2, Q_3 with A_2A_3, A_3A_1, A_1A_2 respectively.

Theorem. *For any triangles $A_1A_2A_3$ and $B_1B_2B_3$,*

$$\frac{\overline{P_1A_2}\cdot\overline{P_2A_3}\cdot\overline{P_3A_1}}{\overline{P_1A_3}\cdot\overline{P_2A_1}\cdot\overline{P_3A_2}} \cdot \frac{\overline{Q_1A_2}\cdot\overline{Q_2A_3}\cdot\overline{Q_3A_1}}{\overline{Q_1A_3}\cdot\overline{Q_2A_1}\cdot\overline{Q_3A_2}} \cdot \frac{\overline{R_1A_2}\cdot\overline{R_2A_3}\cdot\overline{R_3A_1}}{\overline{R_1A_3}\cdot\overline{R_2A_1}\cdot\overline{R_3A_2}} = 1$$

This formula is obtained at once by applying the theorem of Menelaus to each of the transversals $P_1R_2Q_3$, etc., multiplying, and rearranging.

380. Corollary. *Triangles $A_1A_2A_3$ and $B_1B_2B_3$ are in perspective in the sense that A_1B_1, A_2B_2, A_3B_3 are concurrent, if and only if*

$$\frac{\overline{Q_1A_2}\cdot\overline{Q_2A_3}\cdot\overline{Q_3A_1}}{\overline{Q_1A_3}\cdot\overline{Q_2A_1}\cdot\overline{Q_3A_2}} = \frac{\overline{R_1A_3}\cdot\overline{R_2A_1}\cdot\overline{R_3A_2}}{\overline{R_1A_2}\cdot\overline{R_2A_3}\cdot\overline{R_3A_1}}$$

For it is necessary and sufficient that P_1, P_2, P_3 be collinear; that is, that

$$\frac{\overline{P_1A_2}\cdot\overline{P_2A_3}\cdot\overline{P_3A_1}}{\overline{P_1A_3}\cdot\overline{P_2A_1}\cdot\overline{P_3A_2}} = 1$$

If, in the equation of **379**, two of the fractions are equal to 1, the third has the same value, hence the following theorem:

381. Theorem. *If two triangles are in perspective in two ways, they are in perspective in three ways. That is, if A_2B_3, A_3B_1, A_1B_2 are concurrent, and A_3B_2, A_1B_3, A_2B_1 are concurrent, then A_1B_1, A_2B_2, A_3B_3 are concurrent.*

382. The last theorem can be thrown into a somewhat different form, as follows:

Theorem. *Let P and Q be any points in the plane of triangle $A_1A_2A_3$. Let A_2P and A_3Q meet at B_1, A_3P and A_1Q at B_2, A_1P and A_2Q at B_3. Then A_1B_1, A_2B_2, A_3B_3 are concurrent at a point R.*

If A_3Q and A_2P meet at C_1, and so on, the triangles $A_1A_2A_3$ and $C_1C_2C_3$ are likewise, obviously, in perspective. Moreover, it can be proved that $B_1B_2B_3$ and $C_1C_2C_3$ are in perspective, and that these three centers of perspective are collinear. The figure should be drawn in full; it will repay detailed study.

383. If lines AA', BB', CC' meet at a point O, these six points can be assorted into pairs of triangles in four different ways, such as ABC with $A'B'C'$, $A'BC$ with $AB'C'$, and so on. Each pair determines an axis of perspective. Now the six pairs of lines, such as BC and $B'C'$, intersect in six points, and these must lie three by three on the four axes of perspective. In other words:

Theorem. *If AA', BB', CC' are concurrent, the intersections of corresponding connectors, as AB with $A'B'$, AC' with $A'C$, etc., lie by threes on four lines, and therefore are vertices of a complete quadrilateral. Similarly, if three pairs of lines meet at collinear points, the lines through corresponding intersections of these lines are the sides of a complete quadrangle.*

384. Theorem. *Each triangle whose sides are three of the lines of a complete quadrilateral is in perspective with the diagonal triangle of the quadrilateral.*

For the two triangles have as axis of perspective the fourth line of the complete quadrilateral.

Corollary. *Let each point of intersection of two diagonals of a complete quadrilateral be connected with the two remaining vertices of the quadrilateral. The six connectors are concurrent by threes at four points, and therefore are the lines of a complete quadrangle. Thus with every complete quadrilateral there is associated a complete quadrangle having the same diagonal triangle; and conversely. Through each point of the quadrilateral, there is one side of the quadrangle. Each triangle of the quadrangle is in perspective with one triangle of the quadrilateral and the diagonal triangle, the common center of perspective being the fourth vertex of the quadrangle and the axis of perspective the fourth line of the quadrilateral.*

385. Theorem of Pascal. *If a hexagon is inscribed in a circle, the intersections of opposite sides are collinear. That is, if six points, P, Q', R, P', Q, R' lie on a circle in any order, the intersections of PQ' with $P'Q$, QR' with $Q'R$, RP' with $R'P$ are on a line.*

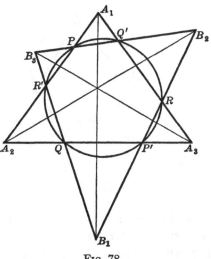

FIG. 78

Let PQ', QR', RP', extended if necessary, be the sides of triangle $B_1B_2B_3$, and $P'Q$, $Q'R$, $R'P$ the sides of triangle $A_1A_2A_3$. Then

$$\overline{Q'A_1}\cdot\overline{RA_1} = \overline{PA_1}\cdot\overline{R'A_1},$$ and similarly for A_2 and A_3.

Multiplying the three equations, and separating,

$$\frac{\overline{QA_2}\cdot\overline{RA_3}\cdot\overline{PA_1}}{\overline{QA_3}\cdot\overline{RA_1}\cdot\overline{PA_2}} = \frac{\overline{P'A_2}\cdot\overline{Q'A_3}\cdot\overline{R'A_1}}{\overline{P'A_3}}\frac{\overline{Q'A_3}\cdot\overline{R'A_1}}{\overline{Q'A_1}\cdot\overline{R'A_2}}$$

whence by direct application of **380**, the triangles $A_1A_2A_3$ and $B_1B_2B_3$ are in perspective. Their axis of perspective is the line whose existence we wished to prove.

386. This famous theorem was discovered by Blaise Pascal in 1640, when he was only sixteen years old. From the point of view of elementary geometry, it is not especially significant, since the converse is not true. It is a special case of the following more general theorem: the intersections of opposite sides of a hexagon are collinear, if and only if the vertices of the hexagon lie on a curve of the class known as conic sections. Only if it is known that five vertices are on a circle, can we assert that the sixth will be on the circle, and the theorem is therefore of limited application. In the field of projective geometry, on the other hand, this theorem has considerable importance.

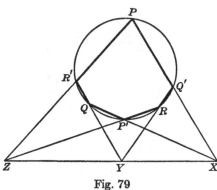

Fig. 79

We will state without proof some further properties of the figure. If six points are marked on a circle, we may draw connectors so as to form a hexagon (not necessarily a convex hexagon) in many different ways; sixty, in fact. Every such hexagon determines a Pascal line, and these lines are easily seen to be all distinct. The fifteen connectors of the six given points intersect at forty-five other points; and through each of these there are four of the sixty Pascal lines. The Pascal lines meet also by threes at twenty other points, called Steiner points, one on each line; and again by threes at sixty other points, Kirkman points, three to a line. The twenty

Steiner points lie by fours on fifteen other lines, and the sixty Kirkman points by threes on twenty other lines. These statements will serve to indicate the almost boundless possibilities in this innocent-looking theorem.*

387. Related to the theorem of Pascal is that of Brianchon (1806):

Theorem. *If a hexagon is circumscribed about a circle, the connectors of opposite vertices are concurrent.*

The simplest proof is based on the method of polar reciprocation (**134**). We consider an auxiliary figure, in which we have found the pole of every line of the given figure, and the polar of every point. In the given figure, six lines are tangent to the circle; in the auxiliary figure, six points lie on the circle. A vertex of the original hexagon yields a side or connector of the new hexagon, and vice versa. The theorem of Pascal for the auxiliary figure, translated back to the given figure, is precisely the theorem of Brianchon.

The theorem can be proved by direct methods, without the use of poles and polars,† but the proof is roundabout and in no way instructive.

Like the theorem of Pascal, that of Brianchon is a special case of a more general theorem, and the converse is not true. It admits of generalizations and extensions parallel to those which we have suggested for the theorem of Pascal.

388. Theorem of Pappus. *If the vertices of a hexagon fall alternately on two lines, the intersections of opposite sides are collinear.*

That is, if PQR and $P'Q'R'$ are two lines, the sides PQ' and $P'Q$, QR' and $Q'R$, RP' and $R'P$ of the hexagon $PQ'RP'QR'$ meet at collinear points.

---

* For full treatment see Lachlan, *l.c.*, p. 113.
† Cf. Lachlan, *l.c.*, p. 116.

For the triangles whose sides are PQ', QR', RP', and $P'Q$, $Q'R$, $R'P$ have two axes of perspective, namely PQR and $P'Q'R'$; they have therefore a third axis of perspective.

This theorem suggests the same sort of extensions as suggested in **386**; and a "dual" set of theorems as well.

389. Some theorems, apparently unrelated to the theorem of Pascal, may be proved by means of it. A few examples may be noted.

Theorem. *From two vertices of a triangle, $A_1A_2A_3$, let intersecting lines A_2P and A_3P be drawn. Let P_2 and P_3 be the feet of the perpendiculars from P on A_1A_3 and A_1A_2 respectively; and X_2, X_3 the feet of the perpendiculars from A_1 on A_2P and A_3P respectively. Then P_2X_2, P_3X_3, and A_2A_3 are concurrent.*

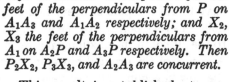

This result is established at once by applying the theorem of Pascal to the hexagon $A_1P_3X_3PX_2P_2A_1$, which is inscribed in the circle drawn on A_1P as diameter.

FIG. 80

390. Theorem. *Given a triangle $A_1A_2A_3$, and a line through a point P, meeting the sides at X_1, X_2, X_3 respectively. Let A_1P meet the circumcircle at R_1; etc. Then X_1R_1, X_2R_2, X_3R_3 are concurrent at a point on the circumcircle.*

For if R_1X_1 meets the circumcircle at T_1, we may apply Pascal's theorem to hexagon $A_1R_1T_1R_2A_2A_3A_1$, and we find that R_2X_2 passes also through T_1. And similarly for R_3X_3.

The converse theorem, which is an immediate consequence, may be thrown into the following form.

391. Theorem. *Let two triangles be inscribed in the same circle and in perspective. The lines connecting any point of the circle to the vertices of one triangle meet the corresponding* *sides of the other triangle in three points that are collinear with the center of perspective.*

392. Theorem. *If P and Q are isogonal conjugates, $P_1P_2P_3$ and $Q_1Q_2Q_3$ their pedal triangles, and if P_2Q_3 and P_3Q_2 meet at X_1, etc.; then X_1, X_2, X_3 lie on PQ.*

For the center of the common pedal circle is at R, the midpoint of PQ; and PP_1 meets Q_1R at a point P_1' on the circle. Applying Pascal's theorem to the hexagon $Q_1P_2P_2'Q_2P_1P_1'$, we find that P, R, X_1 are collinear.

Exercise. In this chapter the following sections contain theorems to be proved by the student: **378, 379, 381, 384, 387, 389, 391.**

CHAPTER XIV

PEDAL TRIANGLES AND CIRCLES

393. In this chapter we consider a remarkable group of theorems concerning the pedal triangles and circles and the nine point circles in the figure determined by four points. These results lead us naturally to a reconsideration of the theorem of Feuerbach, and to some generalizations of that theorem. We end the chapter with a brief account of the orthopole of a line with regard to the triangle.

394. The first group of theorems deals with a complete quadrangle. If four points are given, the pedal triangles of each with regard to the triangle of the other three are similar; the pedal circles of the points and the nine point circles of the triangles, are concurrent at a single point.*

Let four given points A_1, A_2, A_3, A_4 be not concyclic nor orthocentric. We shall denote the foot of the perpendicular from A_1 to A_2A_3 by P_{14}; and so on. The mid-point of the line A_2A_3 shall be M_{23}, and so on. This apparently awkward symbolism will quickly justify itself. The pedal triangle of A_4 with regard to triangle $A_1A_2A_3$ is $P_{41}P_{42}P_{43}$ and the nine point circle of triangle $A_1A_2A_3$ passes through the feet of the altitudes, P_{14}, P_{24}, P_{34} and the mid-points of the sides, M_{23}, M_{31}, M_{12}.

395. Theorem. *The pedal triangles in a complete quadrangle are directly similar in the sense*

$$P_{12}P_{13}P_{14} \sim P_{21}P_{24}P_{23} \sim P_{34}P_{31}P_{32} \sim P_{43}P_{42}P_{41}$$

* The theorems of this section, though undoubtedly of much earlier origin, were assembled in 1912 by Happach, *Zeitschrift für Math. und Nat. Unterricht*, **43**, p. 175. We have modified and shortened the proofs.

This is an immediate consequence of the fundamental Miquel equation (**186**):

$$\angle P_{12}P_{13}P_{14} = \angle A_2A_1A_4 + \angle A_4A_3A_2$$

$$\angle P_{34}P_{31}P_{32} = \angle A_4A_3A_2 + \angle A_2A_1A_4$$

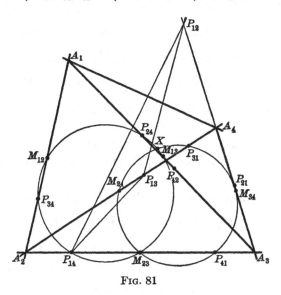

Fig. 81

showing that these angles of two pedal triangles are equal. But in either right hand member we may transpose (**18** f);

$$\angle A_2A_1A_4 + \angle A_4A_3A_2 = \angle A_1A_2A_3 + \angle A_3A_4A_1$$

showing that angles $\angle P_{21}P_{24}P_{23}$ and $\angle P_{43}P_{42}P_{41}$ are also equal to the two just named. Similarly for the other angles of the pedal triangles.

The cyclic symmetry of the arrangement of the subscripts is not immediately obvious. But if we label the vertices of a tetrahedron 1, 2, 3, 4; then if the tetrahedron rests with one vertex upward, corresponding to the point whose pedal triangle is under consideration, the other three vertices will lie in positive order as named in the theorem.

396. Theorem.   *The nine point circles of the four triangles are concurrent.*

Let the circles $M_{12}M_{13}M_{23}$ and $M_{23}M_{24}M_{34}$ meet again at X.

Then $\measuredangle\, M_{12}XM_{23} = \measuredangle\, M_{12}M_{13}M_{23'} = \measuredangle\, A_3A_2A_1$

$\measuredangle\, M_{23}XM_{24} = \measuredangle\, M_{23}M_{34}M_{24} = \measuredangle\, A_4A_2A_3$

Adding, $\measuredangle\, M_{12}XM_{24} = \measuredangle\, A_4A_2A_1 \quad = \measuredangle\, M_{12}M_{14}M_{24}$

so that X lies on the circle through M_{12}, M_{14}, M_{24}, which is the nine point circle of $A_1A_2A_4$.

397. Theorem.   *The pedal circles of each of the four points with regard to the triangle of the other three are also concurrent at X.*

That is, for instance, P_{12}, P_{13}, P_{14}, X are concyclic.

For $\measuredangle\, P_{24}XP_{21} = \measuredangle\, P_{24}XM_{23} + \measuredangle\, M_{23}XP_{21}$

$= \measuredangle\, P_{24}M_{13}M_{23} + \measuredangle\, M_{23}M_{34}P_{21}$

$= \measuredangle\, A_1A_3,\, A_1A_2 + \measuredangle\, A_2A_4,\, A_4A_3$

$= \measuredangle\, A_4A_2A_1 + \measuredangle\, A_1A_3A_4$

$= \measuredangle\, P_{24}P_{23}P_{21}$ \hfill **(395)**

398. Theorem.   *The second intersection of the pedal circles of two points lies on the line joining the other two.*

Consider, for instance, the circles $P_{12}P_{13}P_{14}$ and $P_{43}P_{42}P_{41}$, which intersect at X and at a second point Y_{14}.   Then

$\measuredangle\, P_{14}Y_{14}P_{41} = \measuredangle\, P_{14}Y_{14}X + \measuredangle\, XY_{14}P_{41}$

$= \measuredangle\, P_{14}P_{12}X + \measuredangle\, XP_{42}P_{41}$

$= \measuredangle\, P_{14}P_{12},\, P_{42}P_{41} + \measuredangle\, P_{42}XP_{12}$

Now X is on the nine point circle of $A_1A_3A_4$, and we find that

$$\measuredangle \, P_{42} \, X P_{12} = 2 \measuredangle \, A_1 A_3 A_4 \qquad \qquad (252 \, d)$$

Also

$$\measuredangle \, P_{14} P_{12}, \, P_{42} P_{41} = \measuredangle \, P_{14} P_{12}, \, A_2 A_3 + \measuredangle \, A_2 A_3, \, P_{42} P_{41}$$

$$= 2 \measuredangle \, A_4 A_3 A_1$$

Hence $\qquad \qquad \measuredangle \, P_{14} Y_{14} P_{41} = 0$

and the point Y_{14} lies on $P_{14} P_{41}$, which is the same as $A_2 A_3$.

Corollary. *The pedal triangles $P_{12} P_{13} P_{14}$ and $P_{43} P_{42} P_{41}$ are in perspective at Y_{14}; that is, $P_{12} P_{43}$, $P_{13} P_{42}$, and $P_{14} P_{41}$, pass through Y_{14}.*

For $\qquad \measuredangle \, P_{12} \, Y_{14} P_{14} = \measuredangle \, P_{12} P_{13} P_{14}$

$$\measuredangle \, P_{43} Y_{14} P_{41} = \measuredangle \, P_{43} P_{42} P_{41}$$

and we have proved the right-hand members equal in the similar pedal triangles.

Theorem. *The center of similitude of the pedal triangles is the point X.*

399. Theorem. *The isogonal conjugate of A_4 with regard to $A_1 A_2 A_3$ and that of A_1 with regard to $A_2 A_3 A_4$ lie on the perpendicular to $A_2 A_3$ at Y_{14}, and are equally distant from Y_{14} (231, 236). The quadrangle of the four isogonal conjugates, namely of each point with regard to the triangle of the other three, is similar to that of the circumcenters of the four triangles.*

400. The foregoing theorems require some modification when the four given points are on a circle. Some of the analogous theorems have been proved already; the others can be established without difficulty.

a. *If A_1, A_2, A_3, A_4 are four points on a circle, the segments on the respective Simson lines are congruent in the sense indicated in 395, namely:*

$$\overline{P_{12} P_{13}} = \overline{P_{21} P_{24}} = \overline{P_{34} P_{31}} = \overline{P_{43} P_{42}}, \; etc.$$

b. The four nine point circles are congruent and meet in a point X through which pass the four Simson lines. (Cf. **331**)

c. Any two Simson lines make equal angles with the line joining the other two points.

d. The segments measured from X on the Simson lines are equal by fours,

$$\overline{XP}_{14} = \overline{XP}_{41} = \overline{XP}_{23} = \overline{XP}_{32}, \text{ etc.,}$$

and the twelve feet of the perpendiculars lie by fours on three circles concentric at X (N. Anning). *The connectors $P_{14}P_{41}$, $P_{13}P_{42}$, $P_{12}P_{43}$ are parallel.*

e. The isogonals, with regard to any triangle, of the lines to the fourth point are perpendicular to the Simson line of that point.

401. The theorem of Feuerbach would appear to be an easy consequence of **396**.   The argument in the following form is due to Fontené.*

Let P and Q be isogonal conjugates with regard to a triangle $A_1A_2A_3$, and hence have the same pedal circle.   If this circle meets the nine point circle of the triangle at X and Y, then it is evident that the nine point circles of the four triangles whose vertices are A_1, A_2, A_3, P pass through X, and those of A_1, A_2, A_3, Q through Y.   Now, in particular, let P and Q coincide; then the two sets of triangles are one and the same, and X and Y are coincident.   In other words, the pedal circle of the incenter or an excenter is tangent to the nine point circle.

But this proof is not quite sound.   For we have no simple defense against the objection that possibly in the general case, when P and Q are distinct, all the nine point circles pass through one of the points, say X, while the other is without significance.   More careful study of these problems leads us

* *Nouvelles Annales*, 1905, p. 260.

to some new and valuable theorems. We make a new beginning and approach the problem from a different angle.*

402. Theorem. *If the sides of the pedal triangle of a point P meet the corresponding sides of triangle $O_1O_2O_3$ at X_1, X_2, X_3 respectively, then P_1X_1, P_2X_2, P_3X_3 meet at a point L common to the circles $O_1O_2O_3$ and $P_1P_2P_3$. That is, L is one of the intersections of the nine point circle of $A_1A_2A_3$ and the pedal circle of P.*

For let OP meet the circle on A_1O as diameter at L_1. Let the reflection of P_1 with regard to O_2O_3 be P_1', so that A_1P_1'

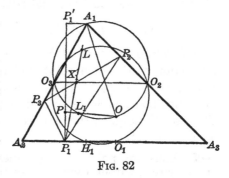

and P_1P_1' are respectively parallel and perpendicular to A_2A_3. Now A_1, O, O_2, O_3, L_1 are concyclic; and also A_1, P, P_2, P_3, L_1 and P_1' lie on a circle drawn on A_1P as diameter. The feet of the perpendiculars from L_1 to A_1P_2, A_1P_3, P_2P_3, O_2O_3

FIG. 82

are collinear; therefore L_1 is on the circumcircle of triangle $O_3P_3X_1$.

We next prove with some difficulty that L_1, X_1, and P_1' are collinear. Since P_1' is on the circle drawn on A_1P as diameter,

$$\angle P_1'L_1P_3 = \angle P_1'PP_3 = \angle O_2O_3,A_1P_3 = \angle X_1O_3P_3$$

But we have shown that O_3, P_3, L_1, X_1 are on a circle; hence

---

* These theorems were given by Fontené, with different proofs (*Nouvelles Annales*, 1905, p. 504; 1906, p. 55); these direct proofs were supplied by Bricard (*ibid.*, 1906, p. 59). The central theorem, **404**, however, has been independently discovered by others: Weill, *ibid.*, 1880, p. 259; McCay, *Transactions of the Irish Royal Academy*, XXIX, p. 310; Griffiths, *Educational Times*, 1857,

$$\angle X_1O_3P_3 = \angle X_1L_1P_3$$

so that
$$\angle P_1'L_1P_3 = \angle X_1L_1P_3$$

and therefore P_1', L_1, X_1 are on a line. We now turn a part of the figure over on O_2O_3 as axis. A_1 turns over to H_1, O_2 and O_3 remain in place; hence the circle through these points, namely the circle on A_1O as diameter, turns over into the nine point circle. P_1' turns over into P_1. Hence if L be the reflection of L_1, it lies on the nine point circle and also on X_1P_1, the reflection of the line $L_1X_1P_1'$.

Next, L lies also on the pedal circle of P, since

$$\overline{X_1P_1}\cdot\overline{X_1L} = \overline{X_1P_1'}\cdot\overline{X_1L_1} = \overline{X_1P_2}\cdot\overline{X_1P_3}$$

Thus if P_2P_3 meets O_2O_3 at X_1, then P_1X_1 passes through an intersection of the nine point and pedal circles. That the three lines P_1X_1, P_2X_2, P_3X_3 pass through the same point of the nine point circle depends directly on **333**; since the line OP passes through O, the orthocenter of $O_1O_2O_3$. This completes the proof.

403. Thus any point P determines a point L on the nine point circle, through which pass the pedal circles and the nine point circles of the quadrangle $A_1A_2A_3P$, according to **396** and **397**. From the proof we see that *two points P and Q actually determine distinct points on the nine point circle, unless they are collinear with the circumcenter O, when they determine the same point L.* This completes the proof of the theorem of Feuerbach above, and at the same time furnishes some generalizations.

404. Theorem. *If a point moves on a fixed line through the circumcenter, its pedal circle passes through a fixed point on the nine point circle.*

405. Theorem. *The pedal circle of a point is tangent to the nine point circle, if and only if the point and its isogonal conjugate lie on a line through the orthocenter.*

Corollary. *The theorem of Feuerbach is merely a special case* of this theorem, since the incenter, or any one of the excenters, is its own isogonal conjugate.

Construction. *If a line through the circumcenter is given, the corresponding point on the nine point circle may be found by drawing the reflections of the line as shown in the proof of* **403**; *or by virtue of the fact that it is the intersection of the Simson lines of the points where the line cuts the circumcircle.*

406. Another approach to these same problems is based on the so-called orthopole, which was extensively studied by Neuberg, Soons, Gallatly, and others. This treatment is given at some length by Gallatly,* and we will content ourselves with a survey of the leading results.

Theorem. *If perpendiculars are dropped on any line from the vertices of a triangle, the perpendiculars to the opposite sides from their feet are concurrent at a point called the orthopole of the point.*
As a line moves parallel to itself, the locus of the orthopole is a line perpendicular to the given line. The orthopole of a line which meets the circumcircle is the point of intersection of the Simson lines of the points where the line cuts the circle. In other words, a Simson line is the locus of the orthopole of the lines through the point. If a line passes through the circumcenter, its orthopole lies on the nine point circle.
If a line cuts the circumcircle at P and Q, the lines from the vertices of the triangle perpendicular to this line may be extended to the circumcircle; from these points perpendiculars are drawn to the opposite sides. These are concurrent at a point R of the circumcircle. From the feet of the perpendiculars to PQ, the perpendiculars to the opposite sides are concurrent at the orthopole S. The Simson lines of P, Q, R all pass through S, which is the orthopole of each of the lines PQ, PR, QR. (Cf. **337** ff.)
The orthopole of a line has the same power with regard to the pedal circles of all points on the line.

(This includes **404** as a special case, when the power is zero.)

* *Modern Geometry of the Triangle,* chap. VI.

CHAPTER XV
SHORTER TOPICS

407. In this chapter we introduce first the physical concepts of center of gravity and resultant of forces, and establish a number of geometric theorems based on these statical methods. We then consider several groups of theorems for triangles and quadrangles, many of which are left to be worked out fully by the reader.

STATICAL THEOREMS

408. The notion of center of gravity is a familiar one, and may be made definite for the purposes of geometry as follows.

Definition. Given a number of points in a plane, and a weight at each; let the weight at P_1 be m_1, and so on. Let two intersecting lines be chosen, and let the distance from P_1 to the first be d_1 and to the second d_1'. Then the point whose distances from these lines are given by

$$d = \frac{m_1d_1 + m_2d_2 + \ldots \ldots + m_nd_n}{m_1 + m_2 + \ldots \ldots + m_n}$$

$$d' = \frac{m_1d_1' + m_2d_2' + \ldots \ldots + m_nd_n'}{m_1 + m_2 + \ldots \ldots + m_n}$$

is called the *center of gravity* of the system.

409. It can be proved at once that the perpendicular from the center of gravity to *any other* line is given by a like formula. In geometric theorems, we usually consider equal weights, and speak of the center of gravity of a set of points as that point whose distance from any line is the average of the distances to the given points from the same line. The center of gravity of a straight line segment is taken as its mid-point, and that of a triangular area at the centroid. In dealing with centers of gravity, any system of weights may be replaced by a single weight equal to their sum and placed at their center of gravity.

410. Theorem.   *The center of gravity of the vertices of a triangle is the median point.*

The proof illustrates well the method.   Let equal unit weights be placed at the vertices of the triangle; then two of them may be replaced by a weight 2 at the point midway between them.   The center of gravity of the system of this weight and the third unit weight is evidently on the median, and trisects it.

411. Theorem.   *The center of gravity of four points constituting an orthocentric system is at the center of the common nine point circle.*

Several easy proofs are obvious, and the theorem suggests the following:

Theorem (Beltrami).   *The center of gravity of the incenter and excenters of a triangle is at the circumcenter.*

Theorem.   *The connectors of mid-points of opposite sides of a quadrangle, and the connector of the mid-points of the diagonals, have a common mid-point, which is the center of gravity of the four points.*

Theorem.   *If a triangle is divided by a median, the triangles formed are equal in area; the center of gravity of the original triangle bisects the line joining the centers of gravity of the half-triangles.*

412. Theorem.   *The center of gravity of the perimeter of a triangle (as when a piece of wire is bent into triangular form) is at the center of the Spieker circle (**364**).*

For the weight of each side may be replaced by a weight at its mid-point, proportional to its length.   Then in triangle $O_1O_2O_3$, we have weights at the vertices proportional to the opposite sides, and the center of gravity is easily seen to be at the incenter of that triangle.

413. Theorem. *If the sides of a triangle are divided proportionally at P_1, P_2, P_3, so that*

$$\frac{\overline{A_2P_1}}{\overline{P_1A_3}} = \frac{\overline{A_3P_2}}{\overline{P_2A_1}} = \frac{\overline{A_1P_2}}{\overline{P_3A_2}} = \frac{p}{q}$$

the center of gravity of triangle $P_1P_2P_3$ is at M. (Cf. **358**)

For let equal weights be placed at P_1, P_2, P_3. Let each be divided into parts proportional to p and q, and these parts placed at the extremities of the corresponding sides. Then the center of gravity of the system is unchanged; but since we now have equal weights at A_1, A_2, and A_3, the center of gravity is at M.

Exercise. Modify this proof to apply to the case that the sides are divided externally.
Similarly:

414. Theorem. *If points are marked on each side of a triangle equidistant from the mid-points, so that*

$$\overline{A_2P_1} = \overline{Q_1A_3}, \ \overline{A_3P_2} = \overline{Q_2A_1}, \ \overline{A_1P_3} = \overline{Q_3A_2}, \ then$$

*(a) triangles $P_1P_2P_3$, $Q_1Q_2Q_3$ are equal in area (**107**).*
(b) the line joining their median points is bisected at M.

415. Theorem.* *If X_1, X_2, X_3 are on the sides A_2A_3, A_3A_1, A_1A_2 of a triangle, so that $\overline{A_2X_1} = \overline{A_3X_2} = \overline{A_1X_3}$, then the locus of the center of gravity of triangle $X_1X_2X_3$ is a line through M.*

For it is easy to show by the methods used above that the centers of gravity of any two such triangles are collinear with M. Various possible generalizations suggest themselves, but we have sufficiently demonstrated the possibilities of the method.

* M. d'Ocagne, *Mathesis*, 1887, p. 265.

416. The process of addition of vectors, or composition of forces by the so-called parallelogram law, suggests a number of interesting geometric theorems. If two forces or velocities are represented by direction and magnitude by lines issuing from a point, their resultant is defined as the quantity represented by the diagonal of the parallelogram having these lines as two of its sides.

Theorem. *Three or more forces at a point have a unique and definite resultant, in whatever order they are combined.*

Theorem (Sylvester). *The resultant of three equal forces OA_1, OA_2, OA_3, acting in any directions at a point O, is the force represented by OH, where H is the orthocenter of triangle $A_1A_2A_3$.*

For with our usual notation, the resultant of OA_2 and OA_3 is $OO_1' = 2\,OO_1$. But OO_1' is equal and parallel to A_1H (**256**), and therefore $OO_1'HA_1$ is parallelogram, whose diagonal OH is the desired resultant. More generally:

Theorem. *If PA_1, PA_2, PA_3 are any three forces in a plane, acting at a point P, and if M is the median point of triangle $A_1A_2A_3$, the resultant of the forces is 3 PM.*

The proof is left as an exercise. This and numerous similar theorems may be found in an article by Alison.*

THE CYCLIC QUADRANGLE

417. It has been established in **265** that if four points A_1, A_2, A_3, A_4 lie on a circle; and if H_1, H_2, H_3, H_4 are the orthocenters of triangles $A_2A_3A_4$, etc., then figures $A_1A_2A_3A_4$ and $H_1H_2H_3H_4$ are congruent, with corresponding sides parallel and in opposite directions, so that A_1H_1, A_2H_2, etc., all have a common mid-point P. Further (**400, 327**), the nine point circles of the four triangles $A_1A_2A_3$, etc., all pass

* "Statical Proofs of Some Geometrical Theorems," *Proceedings of Edinburgh Mathematical Society,* IV, 1886, p. 58.

through P, as does the Simson line of each of the points A_1, A_2, A_3, A_4 with regard to the triangle of the other three. And moreover, since A_1 is the orthocenter of $H_2H_3H_4$, etc., it is evident that the nine point circles and Simson lines of $H_1H_2H_3H_4$ also pass through P. Yet again, A_1 is the orthocenter of each of the triangles $A_2A_3H_4$, $A_2H_3A_4$, $H_2A_3A_4$; and similarly for A_2, A_3, A_4. We may combine all these into the following

> **Theorem.** *Let A_1, A_2, A_3, A_4 be four points on a circle, H_1, H_2, H_3, H_4 the orthocenters of triangles $A_2A_3A_4$, etc. If from the eight points we choose four with different subscripts, three from one set and the fourth from the other, they form an orthocentric system. There are eight such systems. On the other hand, if we choose all the points of one set, or two from each set, with all different subscripts, we have four points on a circle. There are four pairs of such circles; thus the eight points lie by fours on eight equal circles.*

Evidently all these systems have in common the point P. Thus, for instance, the Simson line of any one of the eight points with regard to the triangle of any three others concyclic with it passes through P.

> **418. Theorem** (Weill). *The Simson line of A_4 with regard to triangle $A_1A_2A_3$ is the same as that of H_4 with regard to triangle $H_1A_2A_3$.*

For each passes through P, and also through the intersection of A_4H_1 with A_2A_3.

> **Corollary.** *This same line enacts eight different rôles; namely, it is the Simson line*
>
> *of A_4 with regard to $A_1A_2A_3$; of H_4 with regard to $H_1H_2H_3$*
>
> *of H_1 " " " $A_2A_3H_4$; of A_1 " " " $H_2H_3A_4$*
>
> *of H_2 " " " $A_3H_4A_1$; of A_2 " " " $H_3A_4H_1$*
>
> *of H_3 " " " $H_4A_1A_2$; of A_3 " " " $A_4H_1H_2$*

and thus the thirty-two possible Simson lines in the eight con-cyclic systems fall together by eights into four distinct lines.

419. Theorem.   *Similarly the nine point circles of the eight orthocentric systems are concurrent at P; and, since they are equal, their centers are on a circle with center at P.   These eight centers form a figure similar to that of the A's and H's, in the ratio $1:2$.*

The reader may prove that the pedal circle of any one of the eight points (A) and (H) with regard to the triangle of any three others passes through P.   It may be seen that there are in all 280 such circles; but that actually many of them are not distinct.   After ascertaining how many have been accounted for already, the others should be investigated.

420. Theorem.   *If the trisectors of the angles of a triangle are drawn so that those adjacent to each side intersect, the inter-sections are vertices of an equilateral triangle.*

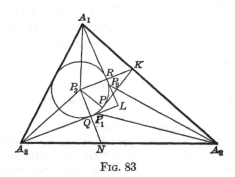

Fig. 83

Let the trisectors adjacent to A_2A_3 be A_2P_1 and A_3P_1, etc.; it is to be proved that $P_1P_2P_3$ is an equilateral triangle.

Extend A_2P_1 and A_1P_2 to meet at L.   Draw the incircle of triangle A_1A_2L, whose center is obviously P_3.   Let Q and R denote its points of contact on LA_2 and LA_1 respectively, and let P_3R meet A_1A_3 at K, and P_3Q meet A_2A_3 at N; let the tangent from K to the circle touch it at P, and meet A_2L at F.

$$\overline{P_3R} = \overline{RK}, \qquad P_3P = \tfrac{1}{2}\,\overline{P_3K},$$

$$\angle PP_3K = 60°, \qquad \angle P_3KP = 30°.$$

Also $\quad \angle QP_3R = 180° - \angle QLR = 120° - \tfrac{2}{3}\,\alpha_3$

Hence $\quad \measuredangle FNQ = \angle FP_3Q = \tfrac{1}{2}\angle QP_3P$

$$= \tfrac{1}{2}(\angle QP_3R - 60°) = 30° - \tfrac{1}{3}\,\alpha_3$$

Again, $\quad \angle P_3NK = \angle P_3KN = \tfrac{1}{2}\angle QLR = 30° + \tfrac{1}{3}\,\alpha_3.$

$$\angle FNK = \tfrac{2}{3}\,\alpha_3, \qquad \angle FKN = \tfrac{1}{3}\,\alpha_3$$

so that F, K, A_3, N are concyclic. Therefore F coincides with P_1, and the tangent to the circle from K passes through P_1.

Similarly the tangent from N to the same circle passes through P_2. Now the figure P_3NK is symmetrical, and NP_2 and KP_1 are symmetrically placed; hence it is easy to see that arc PP_1 equals arc P_2R; but angle PP_3R, and therefore angle $P_1P_3P_2$, equals 60°.

421. This theorem has been generalized by Taylor and Marr.* Their principal result is as follows:

Theorem. *Each angle of a triangle has six trisectors; with each interior angle trisector are associated the two lines making angles of 120° with it. These trisectors intersect in twenty-seven points which lie six by six on nine lines; and these lines are in three sets of parallels, making angles of 60° with one another.*

422. Another theorem of the same type is given by Fuhrmann (*l.c.*, p. 50).

* *Proceedings of Edinburgh Math. Society*, XXXII, 1914, pp. 119–50. The clever proof given above is ascribed by these authors to W. E. Philip.

Theorem. *If four points are on a circle, the incenters of the four triangles form a rectangle whose sides are parallel to the lines connecting the middle points of opposite arcs; and these connectors pass through the center of the rectangle.*

Let A, B, C, D lie on a circle in order, and let a, b, c, d be the incenters of triangles BCD, CDA, DAB, ABC; let the mid-points of arcs AB, BC, CD, DA be M, N, P, Q.

Now the circle with center Q, radius $QA = QD$, passes through b and c (**292**); and also DcM, AbP are straight lines.

Hence $\angle bAD = \angle bcD$

But $\angle bAD = \angle PAD = \angle PMD$

hence $\angle bcD = \angle PMD$, and bc is parallel to MP.

Thus bc and ad are parallel to MP, and ab and cd to NQ. But these are perpendicular, and therefore $abcd$ is a rectangle. Moreover, NQ is perpendicular to a chord bc of a circle whose center is Q, and therefore bisects the chord bc. Thus the intersection of the lines MP and NQ is the center of the rectangle.

More generally:

Theorem. *The sixteen incenters and excenters of the triangles whose vertices are four points on a circle, are the intersections of two sets of four parallel lines, mutually perpendicular.*

423. The following theorems were stated by Steiner without proof; and we shall follow this august example. Part of the proof offers no difficulty, but the complete proof (as supplied by Mention) is long and difficult.*

Theorem. *In a complete quadrilateral the bisectors of the angles are concurrent at sixteen points, the incenters and excenters of the four triangles. These points are the intersec-*

---

* Steiner, *Collected Works*, I, p. 223; Mention, *Nouvelles Annales*, 1862, p. 16; p. 65.

tions of two sets of four circles each, which are members of conjugate coaxal systems. The axes of these systems intersect at the point common to the circumcircles of the quadrilateral.

CIRCLES OF DROZ-FARNY

424. The first of the following theorems was given by Steiner, as usual without proof. The proof, and the extensions which follow, are due to Droz-Farny (*Mathesis*, 1901, p. 22).

Theorem. *If any circle with center at H cuts the lines O_2O_3, O_3O_1, O_1O_2, at P_1, Q_1, P_2, Q_2, P_3, Q_3 respectively, then*

$$\overline{A_1P_1} = \overline{A_2P_2} = \overline{A_3P_3} = \overline{A_1Q_1} = \overline{A_2Q_2} = \overline{A_3Q_3}$$

Let X be any point on O_2O_3, and T_1 the mid-point of A_1H_1. Then we have

$$\overline{A_1X}^2 = \overline{A_1H}^2 + \overline{HX}^2 + 2\,\overline{A_1H}\cdot\overline{HT_1}$$
$$= \overline{HX}^2 + \overline{A_1H}\,(\overline{A_1H} + 2\,\overline{HT_1})$$
$$= \overline{HX}^2 + \overline{A_1H}\cdot\overline{HH_1}$$

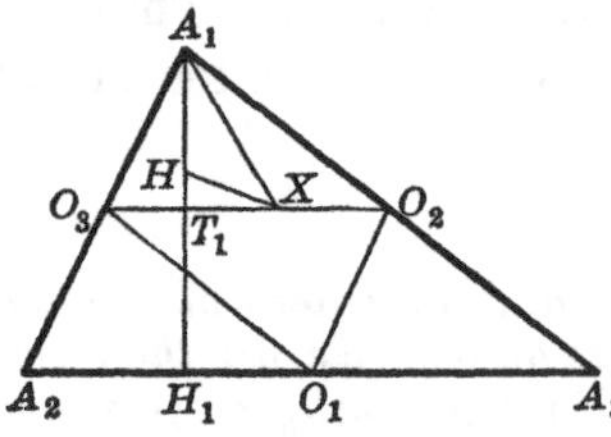

But
$$\overline{A_1H}\cdot\overline{HH_1} = \overline{A_2H}\cdot\overline{HH_2}$$
$$= \overline{A_3H}\cdot\overline{HH_3}$$

FIG. 84

Hence points on O_2O_3, O_3O_1, O_1O_2, equidistant from H, are also equidistant from A_1, A_2, A_3 respectively; and conversely.

425. Theorem. *Conversely, if equal circles are drawn about the vertices of a triangle, they cut the lines joining mid-points of the corresponding sides in six points lying on a circle whose center is at the orthocenter.*

Corollaries. *a. If r is the radius of the equal circles about A_1, A_2, A_3, and R_0 the radius of the circle about H, then (cf. **255**)*

$$R_0^2 = 4R^2 + r^2 - \tfrac{1}{2}(a_1^2 + a_2^2 + a_3^2)$$

b. If circles equal to the circumcircle are drawn about the ver-

tices of a triangle, they cut the lines joining mid-points of the adjacent sides in points of a circle with center H, and radius

$$R_0^2 = 5\,R^2 - \tfrac{1}{2}\,(a_1^2 + a_2^2 + a_3^2)$$

426. *a.* **Theorem.** *Let circles be drawn with centers at the feet of the altitudes, and passing through the circumcenter; they cut the corresponding sides in six points on a circle whose center is H.*

For if, as usual, F denotes the mid-point of OH, the center of the nine point circle, and if the circle with center H_1 and radius H_1O cuts A_2A_3 at P_1 and P_1', then

$$\overline{HP_1}^2 = \overline{HH_1}^2 + \overline{H_1O}^2 = 2\,\overline{H_1F}^2 + \tfrac{1}{2}\overline{OH}^2 \qquad \textbf{(96)}$$

and since $\overline{H_1F} = \overline{H_2F} = \overline{H_3F}$, we have $\overline{HP_1} = \overline{HP_2} = \overline{HP_3}$.

b. **Theorem.** *Circles about the mid-points of the sides, and passing through H, cut the sides in six points lying on a circle whose center is O, equal to the circle of the foregoing theorem.*

For let S_1 lie on A_2A_3, so that $\overline{O_1S_1}$ equals $\overline{O_1H}$; then we have

$$\overline{OS_1}^2 = \overline{OO_1}^2 + \overline{O_1S_1}^2 = \overline{OO_1}^2 + \overline{O_1H}^2$$
$$= \overline{OO_1}^2 + \overline{O_1H_1}^2 + \overline{HH_1}^2 = \overline{H_1O}^2 + \overline{H_1H}^2 = \overline{HP_1}^2$$

c. **Corollary.** *The circles of the two foregoing theorems are equal to that of* **425** *b.*

For $\qquad \overline{OS_1}^2 = \overline{OO_1}^2 + \overline{O_1H}^2$
$$= \overline{OO_1}^2 + \tfrac{1}{4}\,(2\,\overline{A_2H}^2 + 2\,\overline{A_3H}^2 - \overline{A_2A_3}^2) \qquad \textbf{(96 a)}$$

But $\overline{A_2H}^2 = 4\,\overline{OO_2}^2 = 4\,(R^2 - \overline{A_1O}^2) = 4\,(R^2 - \tfrac{1}{4}a_2^2)$, etc.,

whence we get on substituting,

$$\overline{OS_1}^2 = 5\,R^2 - \tfrac{1}{2}\,(a_1^2 + a_2^2 + a_3^2), \text{ as before.}$$

d. **Theorem.** *The circle whose center is H, and radius*

$$\sqrt{5\,R^2 - \tfrac{1}{2}(a_1^2 + a_2^2 + a_3^3)}$$

passes through twelve notable points, two on each of the sides and two on each of the lines joining mid-points of the sides.

427. By a generalization of the foregoing, we have:

Theorem. *Taking as centers the feet of the perpendiculars from a point to the sides of a triangle, circles are drawn passing through the isogonal conjugate of the point. These circles cut the sides on which they are drawn, in six points lying on a circle whose center is the given point; and the circles thus drawn about a pair of isogonal conjugate points are equal.*

PARALOGIC TRIANGLES

428. Theorem. *At the points where a line cuts the sides of a triangle $A_1A_2A_3$, perpendiculars to the sides are drawn, forming a triangle $B_1B_2B_3$ similar to the given triangle. The two triangles are also in perspective; one point of intersection of their circumcircles is the center of similitude, the other is the center of perspective. The circumcircles meet orthogonally.*

The triangles are obviously similar, since homologous angles are equal. They are in perspective because corresponding lines meet at collinear points; that is, the given line is the axis of perspective. Let this line be $X_1X_2X_3$, so that A_2A_3 and B_2B_3 meet at X_1, etc.; and let A_2B_2 meet A_3B_3 at P. Then

$$\measuredangle X_1A_2P = \measuredangle X_1X_3B_2, \quad \measuredangle PA_3X_1 = \measuredangle B_3X_2X_1$$

$$\measuredangle A_2PA_3 = \measuredangle X_3B_1X_2 = \measuredangle A_2A_1A_3$$

and P lies on the circle $A_1A_2A_3$. Similarly P is also on the B-circle.

Again, consider the Miquel point of $X_1X_2X_3$ with regard to either of the triangles. The circle $A_1X_2X_3$ is evidently the same as the circle $B_1X_2X_3$; hence the two triads of Miquel circles coincide, the Miquel point must be common to the two triangles, and it lies on each of the circumcircles. It is obvious that this point is the center of similitude of the triangles.

Generalization. This theorem and proof can be at once extended to the case that lines are drawn from collinear points X_1, X_2, X_3 on the sides of a triangle, making any equal angles with these sides. A second triangle is formed, similar to the first; and the angle between the circumcircles equals the given angle. For a given line $X_1X_2X_3$, but with varying angles, the center of similitude is always the same, but the center of perspective moves along the circle $A_1A_2A_3$. On the other hand, if $X_1X_2X_3$ is moved parallel to itself, and perpendiculars are drawn for each of its positions, the figure $A_1X_2X_3B_1$ will be of constant form, the locus of B_1 will be a line through A_1; hence P is a fixed point, and B_1, B_2, B_3 lie on A_1P, A_2P, A_3P respectively. It follows easily that the Simson line of P is parallel to $X_1X_2X_3$.

From these theorems, first noticed by Steiner, some quite elaborate results have been developed; for instance, the theorem of Sondat that the axis of perspective bisects the line joining the two orthocenters.*

429. The following theorems, due to A. Gob,† are left as exercises. As the theorems are stated, it is assumed that the triangle is acute; some slight modification is necessary for an obtuse triangle, while for a right triangle the theorems have no meaning. Some of these results have already been noticed.

Let the tangents to the circumcircle of triangle $A_1A_2A_3$ form triangle $P_1P_2P_3$ whose incircle is the circle $A_1A_2A_3$.

a. Triangles $P_1P_2P_3$ and $H_1H_2H_3$ are homothetic; and in this similitude, O corresponds to H.

b. Therefore the center of similitude X lies on the Euler line, and

$$\frac{\overline{XH}}{\overline{XO}} = \frac{\overline{HH_1}}{\overline{OP_1}} = \frac{2\,R\,cos\,\alpha_2\,cos\,\alpha_3}{R\,sec\,\alpha_1} = 2\,cos\,\alpha_1\,cos\,\alpha_2\,cos\,\alpha_3$$

* See further Simon, *l.c.*, p. 172, for references.
† *Mathesis*, 1889 supplement.

c. The circumcenter of triangle $P_1P_2P_3$, denoted by O', also lies on the Euler line. (Cf. **315**.)

For it is homologous to F, the circumcenter of $H_1H_2H_3$.

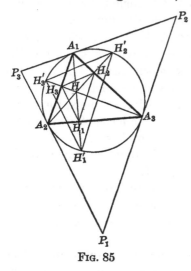

FIG. 85

d. Triangles $P_1P_2P_3$ and $H_1'H_2'H_3'$ are homothetic, and their homothetic center Y lies on OH.

e. The excenters of triangle $P_1P_2P_3$ are vertices of a triangle $Q_1Q_2Q_3$ whose sides are parallel to those of the given triangle; O is its orthocenter, and the center of similitude is X.

f. The pedal triangle of H with regard to $H_1H_2H_3$ is homothetic to $A_1A_2A_3$. The center of similitude is at the same point X.

We get many additional theorems by reversing these results. We may take, for instance, $P_1P_2P_3$ as fundamental triangle, and the results stated yield interesting properties of the incircle. Again, it is especially interesting to take $H_1H_2H_3$ as fundamental triangle. Then H is the incenter, and $P_1P_2P_3$ is formed by tangents to the circumcircle at the middle points of the arcs. The possibilities of the figure will be more fully apparent after extensive study.

430. Power of a triangle. The following curious relations are due to a Spanish geometer, Duran Loriga.*

Definitions. We define the *total power* of a triangle as half the sum of the squares of the sides,

$$P = \tfrac{1}{2}(a_1^2 + a_2^2 + a_3^2)$$

* *Mathesis*, 1895, p. 85.

and the *partial power* of a triangle with regard to any vertex as

$$p_1 = \tfrac{1}{2}(a_2{}^2 + a_3{}^2 - a_1{}^2)$$

Theorems. *a.* $p_1 = a_2 a_3 \cos \alpha_1$

b. $P = p_1 + p_2 + p_3$

c. $P^2 + p_1{}^2 + p_2{}^2 + p_3{}^2 = a_1{}^4 + a_2{}^4 + a_3{}^4$

d. $\Delta = \tfrac{1}{2}\sqrt{p_2 p_3 + p_3 p_1 + p_3 p_1}$

e. p_1 is the power of A_1 with regard to the circle on $A_2 A_3$ as diameter; or with regard to the circle on either $A_2 H$ or $A_3 H$ as diameter; $p_1 = \overline{A_1 H_2} \cdot \overline{A_1 A_3}$.

f. $\dfrac{a_1 p_1}{\cos \alpha_1} = a_1 a_2 a_3 = 4\,\Delta R$

g. $p_1 \tan \alpha_1 = p_2 \tan \alpha_2 = p_3 \tan \alpha_3$

h. If a side of the triangle is given, and the value of any one of these powers, the locus of the third vertex is a circle or a straight line.

431. The following properties of the triangle, due to Schroter, are proved in detail by Fuhrmann. Most of the proof is decidedly difficult unless the methods of projective geometry are used; but it is recommended that the reader draw the complete figure, check all the statements, and try to discover additional relations. The resources of the figure seem well-nigh inexhaustible.

In triangle $A_1 A_2 A_3$, let $O_2 H_3$ meet $O_3 H_2$ at X_1, etc.; let $O_2 O_3$ meet $H_2 H_3$ at Y_1, etc.; let $A_2 A_3$ meet $Y_2 Y_3$ at Z_1, etc. Then:

X_1, X_2, and X_3 lie on the Euler line OH.   (Cf. **392.**)

$A_1 Y_1$, $A_2 Y_2$, $A_3 Y_3$ are parallel to each other and perpendicular to OH.

A_1, X_1, Y_2, Y_3 are collinear, etc.

FY_1 is perpendicular to $Y_2 Y_3$, etc.

O_1Y_1, O_2Y_2, O_3Y_3 *are concurrent at a point* P *of the nine point circle.*

H_1Y_1, H_2Y_2, H_3Y_3 *are concurrent at a point* P' *on the nine point circle.*

The points Z_1, Z_2, Z_3 *lie on the line* PP'.

If H_1P *and* O_1P' *meet at* V_1, *etc., then* Y_1V_1, Y_2V_2, Y_3V_3 *are concurrent at a point of the line* PP'. *Also* V_1 *lies on* A_1X_1, *etc.*

X_1, V_2, V_3 *are collinear; etc.*

Exercise. The following sections of this chapter furnish opportunity for original proofs by the reader: **409, 411, 414, 415, 416, 418, 419, (421), (423), 425, 427, (428), 429, 430, (431)**. The complete proofs of the sections given in parentheses are perhaps more difficult than the others.

CHAPTER XVI

THE BROCARD CONFIGURATION

432. The geometry which constitutes the subject matter of this and the following chapters is relatively modern, being almost entirely the product of the past fifty years. The structure, which is to a considerable extent independent of that which we have already erected, is founded on the properties of two remarkable points related to a triangle, called the Brocard points. These were first noticed as early as 1816, by Crelle, and at about the same time Jacobi and other prominent mathematicians discovered some of their properties. Interest in these researches, however, was not sustained, and the results were soon forgotten.

In 1875 the study of the triangle received an invigorating impulse by the rediscovery by H. Brocard, a French army officer, of the points which bear his name. They attracted more general attention and interest at this time, and it has been estimated that before 1895 over six hundred studies of this field of geometry were published in Europe. Among the prominent investigators were Brocard, Neuberg, Lemoine, McCay, and Tucker; and it is gratifying that each of these names is perpetuated by association with some notable circle or line in the triangle.*

* Perhaps the most satisfactory treatise is that of Emmerich, *Die Brocard'schen Gebilde* (Berlin, 1891). This is practically a compendium of the Brocardian geometry of those loci which are straight lines and circles; it has further a brief but valuable bibliography, and important historical notes.

Another text which treats the subject at length is that of Fuhrmann already frequently cited, *Synthetische Beweise Planimetrischer Sätze* (Berlin, 1890). This treatment is divided into two parts; in addition to the points, lines, and circles associated with the Brocard points, there is a study of some loci which are not elementary and are best approached by analytic methods.

Another analytic treatment of many of these topics is to be found in Casey's *Treatise on the Analytical Geometry of the Point, Line, Circle, and Conic Sections* (Longmans, Green, 1885).

In this chapter we study the various points, lines, and circles which are associated with the geometry of Brocard. We take first the properties of the Brocard points themselves, a pair of isogonal conjugate points lying on the circle which has as diameter the segment from the circumcenter to the symmedian point. On this circle, the Brocard circle, lie also the vertices of two remarkable triangles, also known by the name of Brocard. Some attention is given to a set of circles called the Tucker circles, which are closely related to the Brocard points, and one of which is their common pedal circle. Brief mention is made of the Tarry and Steiner points, followed by consideration of various other triangles simply related to the given triangle, and having the same Brocard angle.

THE BROCARD POINTS

433. Theorem. *In any triangle $A_1A_2A_3$ there is one and only one point Ω, such that*

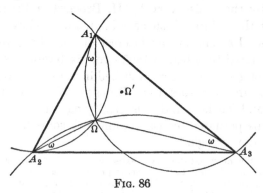

FIG. 86

$$\angle \Omega A_1A_2 = \angle \Omega A_2A_3 = \angle \Omega A_3A_1 = \omega$$

and one point Ω', such that

$$\angle \Omega' A_2A_1 = \angle \Omega' A_3A_2 = \angle \Omega' A_1A_3 = \omega'$$

These two points are called the Brocard points of the triangle.

If a point Ω exists, consider the circle $A_1A_2\Omega$; since angle $\Omega A_2 A_3$ equals the angle $\Omega A_1 A_2$ inscribed in arc ΩA_2, it follows that $A_2 A_3$ is tangent to the circle. In other words the point Ω is common to three known circles, each tangent to a side of the triangle at a vertex, and passing through a second vertex.

Denoting by c_1 the circle tangent to $A_1 A_2$ at A_1, and passing through A_3, etc., the circles c_1, c_2, c_3 are at once proved to be concurrent; this is, in fact, a limiting case of the theorem of Miquel. The point of intersection is necessarily inside the given triangle, and the point Ω is thus completely determined.

Similarly, Ω' is the intersection of three circles c_1', c_2', c_3', where c_1', touches $A_1 A_3$ at A_1 and passes through A_2.

Corollary. $\angle A_2\Omega A_3 = 180° - \alpha_3, \quad \measuredangle A_2\Omega A_3 = \measuredangle A_2 A_3 A_1$

$$\angle A_2\Omega' A_3 = 180° - \alpha_2, \quad \measuredangle A_3\Omega' A_2 = \measuredangle A_3 A_2 A_1$$

Problem. *To construct either Brocard point of a given triangle.*

First solution: draw the circles c_1, c_2, c_3, c_1', c_2', c_3'.

Second solution: if A_1P is parallel to $A_2 A_3$ (**277**) and A_3P is tangent to the circumcircle (**344**) then the circle $A_1 A_3 P$ is c_1, and its intersection with $A_2 P$ is Ω (Fig. 87).

For $\measuredangle A_1 A_3 P = \measuredangle A_1 A_2 A_3, \quad \measuredangle P A_1 A_3 = \measuredangle A_2 A_3 A_1$

whence $\measuredangle A_3 P A_1 = \measuredangle A_3 A_1 A_2 = \measuredangle A_3\Omega A_1$

and by the corollary above, Ω is on the circle $A_1 A_3 P$. Further, by definition,

$$\measuredangle \Omega A_2 A_3 = \measuredangle \Omega A_3 A_1 = \measuredangle \Omega P A_1,$$

and A_2, Ω, P are collinear. A similar construction yields the second point Ω'; from these constructions we deduce a formula of fundamental importance.

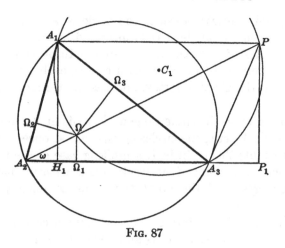

FIG. 87

434. Theorem.

$$cot\ \omega = cot\ \omega' = cot\ \alpha_1 + cot\ \alpha_2 + cot\ \alpha_3$$

Let H_1 and P_1 be the feet of the perpendiculars to A_2A_3 from A_1 and P respectively. Since $\angle PA_2A_3 = \omega$

$$cot\ \omega = \frac{\overline{A_2P_1}}{\overline{PP_1}} = \frac{\overline{A_2H_1}}{\overline{A_1H_1}} + \frac{\overline{H_1A_3}}{\overline{A_1H_1}} + \frac{\overline{A_3P_1}}{\overline{PP_1}}$$

$$= cot\ A_1A_2H_1 + cot\ A_1A_3H_1 + cot\ PA_3P_1$$

$$= cot\ \alpha_2 + cot\ \alpha_3 + cot\ \alpha_1$$

Theorem. *The Brocard points are isogonal conjugates.*

We distinguish Ω and Ω' as the positive and negative Brocard points respectively; ω is called the Brocard angle, and the lines $A_1\Omega$, $A_1\Omega'$, etc., will be called Brocard rays.

435. Theorem. $cot\ \omega = \dfrac{a_1^2 + a_2^2 + a_3^2}{4\ \Delta}$ **(15 g)**

The following relations can be established by trigonometric methods. The similarity of the formula just given to that of **15** g suggests what will be more clear as we progress, that the Brocard angle of a triangle takes rank as of equal importance with the angles of the triangle themselves.

a.
$$cot \ \omega = \frac{1 + cos \ \alpha_1 \ cos \ \alpha_2 \ cos \ \alpha_3}{sin \ \alpha_1 \ sin \ \alpha_2 \ sin \ \alpha_3}$$

$$= \frac{sin^2 \ \alpha_1 + sin^2 \ \alpha_2 + sin^2 \ \alpha_3}{2 \ sin \ \alpha_1 \ sin \ \alpha_2 \ sin \ \alpha_3}$$

$$= \frac{a_1 \ sin \ \alpha_1 + a_2 \ sin \ \alpha_2 + a_3 \ sin \ \alpha_3}{a_1 \ cos \ \alpha_1 + a_2 \ cos \ \alpha_2 + a_3 \ cos \ \alpha_3}$$

b.
$$csc^2 \ \omega = csc^2 \ \alpha_1 + csc^2 \ \alpha_2 + csc^2 \ \alpha_3$$

c.
$$sin \ \omega = \frac{2 \ \Delta}{\sqrt{a_2{}^2 a_3{}^2 + a_3{}^2 a_1{}^2 + a_1{}^2 a_2{}^2}}$$

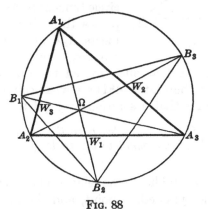

Fɪɢ. 88

436. If we denote the intersections of $A_1\Omega$ and $A_1\Omega'$ with A_2A_3 by W_1 and W'_1 respectively, we easily derive the following useful results.

a.
$$\angle A_1\Omega W_3 = \alpha_1, \ \angle W_3\Omega A_2 = \alpha_3, \ \angle A_2\Omega W_1 = \alpha_2$$

b.
$$\overline{A_2\Omega} = \frac{a_3}{sin \ \alpha_2} \ sin \ \omega$$

c. $$\frac{\overline{A_2\Omega}}{\overline{A_3\Omega}} = \frac{a_3{}^2}{a_1 a_2} = \frac{sin\,(\alpha_3 - \omega)}{sin\,\omega}$$ (15 b)

d. $$\frac{\overline{W_3 A_1}}{\overline{W_3 A_2}} = \frac{a_2\,sin\,\omega}{a_1\,sin\,(\alpha_3 - \omega)} = \left(\frac{a_2}{a_3}\right)^2$$ (84)

437. Problem.   *To construct a triangle, given one side, an adjacent angle, and the Brocard angle.*

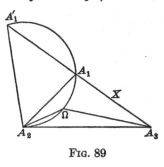

FIG. 89

If α_3 and ω are the given angles, $A_2 A_3$ the given side, we draw triangle $A_2 A_3 \Omega$ with base angles ω and $\alpha_3 - \omega$; and lay off $A_2 A_3 X$ equal to α_3. The required point A_1 lies on $A_3 X$, and also on a circle tangent to $A_2 A_3$ at A_2 which passes through Ω. If this circle meets $A_3 X$ at A_1 and A_1', there are two solutions $A_1 A_2 A_3$ and $A_1' A_2 A_3$, which are at once proved to be inversely similar triangles. We shall later (**481**) see what conditions must be satisfied in order that a solution may exist.

Theorem.   *If two triangles have the same Brocard angle, and an angle of one equal to an angle of the other, they are similar.*

438. We now establish some relations between the Brocard points and the symmedian point, and the properties of the pedal triangles of the former.

Theorem.   *The distances from the symmedian point K to the sides of the triangle are*

$$\overline{KK_1} = \frac{a_1\,tan\,\omega}{2},\ etc.$$ (342, 435)

439. Theorem.   *A Brocard ray, a median, and a symmedian are concurrent. Specifically, $A_1\Omega$, $A_2 K$, and $A_3 M$ meet at a* point; similarly, $A_1\Omega'$, A_2M, and A_3K meet at a point, and the two points are isogonal conjugates. (**344, 436** d, **214**)

Theorem. *Similarly, a Brocard ray, an exmedian, and an exsymmedian are concurrent;* cf. **433**, *problem.*

440. Theorem. *If W_1, W_2, W_3, W_1', W_2', W_3', are the feet of the Brocard rays, and V_1, V_2, V_3 those of the symmedians, then W_1V_2 is parallel to A_1A_2, and $W_1'V_3$ to A_1A_3, etc.*

441. Theorem. *The pedal triangles of Ω and Ω' are similar to the given triangle in the sense*

$$A_1A_2A_3 \sim \Omega_3\Omega_1\Omega_2 \sim \Omega_2'\Omega_3'\Omega_1'$$

The respective centers of similitude are Ω and Ω'; the angles of similitude are $90° - \omega$ and $\omega - 90°$ respectively, and the ratio of similitude in each case $\sin \omega$.

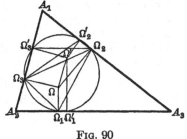

FIG. 90

For since A_1, Ω, Ω_2, Ω_3 are on a circle, we prove by equal angles that triangle $\Omega\Omega_2\Omega_3$ is similar to triangle ΩA_3A_1, and so on. Thus $A_1A_2A_3$ and $\Omega_1\Omega_2\Omega_3$ are composed of similar triangles similarly placed, with Ω self-homologous. The ratio of similitude and the angle of similitude are given by the homologous lines $A_1\Omega$ and $\Omega_3\Omega$.

Theorem. *The pedal triangles of Ω and Ω' are congruent in the sense $\Omega_3\Omega_1\Omega_2 = \Omega_2'\Omega_3'\Omega_1'$.*

442. Theorem. *Conversely, if a triangle $P_1P_2P_3$ inscribed in a given triangle $A_1A_2A_3$ is similar to it in the sense $P_3P_1P_2 \sim A_1A_2A_3$, the Miquel point is at Ω and is the center of similitude; and similarly for Ω'.*

443. Theorem. $\Omega_2\Omega_3'$ *is parallel to A_2A_3, and $\Omega_2'\Omega_3$ is antiparallel to A_2A_3.*

These are proved by means of equal inscribed angles in the common pedal circle.

444. Theorem.   *The radius of the pedal circle of Ω and Ω' is $R \sin \omega$, and its center Q is the mid-point of $\Omega\Omega'$.*

445. Theorem.   *The triangle $O\Omega\Omega'$ is isosceles,*

$$O\Omega = O\Omega', \qquad \angle \Omega O\Omega' = 2\,\omega$$

For in the similar figures $A_1A_2A_3$ and $\Omega_1\Omega_2\Omega_3$, O and Q are homologous points, and $\Omega O Q$ is a right triangle with angle O equal to ω.   Similarly for triangle $\Omega'OQ$.

446.   Further results may be obtained by the use of an auxiliary triangle.

Theorem.   *If the Brocard rays $A_1\Omega$, $A_2\Omega$, $A_3\Omega$ meet the circumcircle again at B_2, B_3, B_1 respectively, the triangles $A_1A_2A_3$ and $B_1B_2B_3$ are congruent. Ω is the negative Brocard point of $B_1B_2B_3$.   Further $\Omega O\Omega'$ is an isosceles triangle, and $\angle \Omega'O\Omega$ is a positive angle $2\,\omega$.*

For (figure 88) the arcs A_1B_1, A_2B_2, A_3B_3 are equal and in the same direction, each subtending the positive angle $2\,\omega$ at O.

And $\qquad \angle\, \Omega B_1 B_3 = \angle\, A_3 A_2 \Omega = \omega.$

447.   *a. Each of the triangles whose bases are sides of the hexagon $A_1B_1A_2B_2A_3B_3$, and whose common vertex is Ω is similar to the given triangle:*

$$A_1A_2A_3 \sim \Omega B_1A_1 \sim B_1A_2\Omega \sim A_2\Omega B_2, \text{ etc.}$$

b. The power of Ω with regard to the circumcircle is

$$\overline{A_1\Omega} \cdot \overline{B_2\Omega} = \overline{A_1B_1} \cdot \overline{A_2B_2} = \overline{A_1B_1}^2 = (2\,R \sin \omega)^2.$$

c. Hence $\quad \overline{O\Omega} = \overline{O\Omega'} = R\sqrt{1 - 4\,sin^2\,\omega}$

$$\Omega\Omega' = 2\,R \sin \omega \sqrt{1 - 4\,sin^2\,\omega}$$

d. The Brocard angle of a triangle is never greater than $30°$. If it has this value, the triangle is equilateral.

THE TUCKER CIRCLES

448. We now investigate a remarkable system of circles, namely the circles circumscribed about the Miquel triangles of the Brocard points (**187, 240, 442**). These are found to have some interesting properties, and several of them are worthy of special attention. At the same time they bring us fuller knowledge of the Brocard points.

449. The *Cosine circle* or *Second Lemoine circle* is defined by the following theorem:

Theorem. *If lines are drawn through the symmedian point K, antiparallel to the sides, their extremities lie on a circle whose center is at K.*

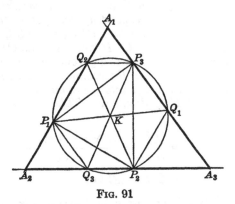

FIG. 91

We know that K bisects any line antiparallel to a side; hence if the line antiparallel to a_1 meets a_3 at P_1 and a_2 at Q_1, etc., we have $\overline{KP_1} = \overline{KQ_1}$. But, moreover, triangle KP_2Q_3 is isosceles with base angles α_1; thus the six distances from K are equal.

Corollaries. Q_2P_3 *is parallel to* A_2A_3, *and* Q_2P_3 *and* P_2Q_3 *are equal. The chords* P_2Q_3, P_3Q_1, P_1Q_2 *are proportional to the cosines of the angles of* $A_1A_2A_3$; *hence the name cosine circle.* P_2P_3 *is perpendicular to* A_2A_3; *hence triangles* $P_1P_2P_3$

and $A_1A_2A_3$ are *directly similar, with corresponding sides perpendicular. Hence the Miquel point of $P_1P_2P_3$ is at Ω, which is also the positive Brocard point of $P_1P_2P_3$ and the center of similitude.* (This can be proved directly by drawing the circles c_1, c_2, c_3 in triangle $P_1P_2P_3$.) *The ratio of similitude of $P_1P_2P_3$ and $A_1A_2A_3$ is* tan ω. *In these similar figures, the circumcenters O and K are homologous. Hence $KO\Omega$ is a right triangle;*

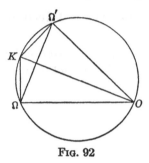

Fig. 92

$$\angle O\Omega K = 90°, \quad \angle KO\Omega = \omega;$$

and the radius of the cosine circle is R tan ω.

450. Theorem. *Similarly, the triangles $Q_1Q_2Q_3$ and $A_1A_2A_3$ are similar, with Ω' as center of similitude and corresponding sides perpendicular. Triangles $P_1P_2P_3$ and $Q_1Q_2Q_3$ are congruent. Hence $OK\Omega$ and $OK\Omega'$ are symmetrically congruent right triangles, on the common hypotenuse OK. In other words, the Brocard points are symmetric as to OK on the circle drawn on OK as diameter. This is called the Brocard circle.*

$$\overline{OK} = \frac{\overline{O\Omega}}{\cos \omega} = \frac{R \sqrt{1 - 4 \sin^2 \omega}}{\cos \omega}$$

$$\overline{\Omega K} = \overline{\Omega' K} = \overline{\Omega O} \tan \omega$$

Corollary. *If W' is the negative Brocard point of $P_1P_2P_3$, W the positive one of $Q_1Q_2Q_3$, and Y, Z the respective symmedian points of these triangles, YZ is parallel to $\Omega\Omega'$, K is the midpoint of YZ, and $WW'\Omega'\Omega$ is a rectangle whose center is K.*

451. K is the center of each of three rectangles inscribed in the triangle, such as $P_2Q_3Q_2P_3$. If rectangles are inscribed in a triangle with their bases in one side, the locus of their centers is a straight line. Hence *the line joining the mid-point of any side to the mid-point of the altitude on that side passes through the symmedian point K.*

This suggests another simple construction for K.   Still another is based on the converse of the principal theorem, namely:

452. Theorem. *If three diameters of a circle terminate in the sides of a triangle, the circle is the cosine circle of the triangle, and its center is the symmedian point K.*

Hence to draw a triangle and its symmedian point simultaneously, we draw any three diameters P_1Q_1, P_2Q_2, P_3Q_3 of a circle; and draw P_2Q_3, P_3Q_1, P_1Q_2 as the sides of triangle $A_1A_2A_3$.   Then the given circle is the cosine circle of $A_1A_2A_3$, and its center is the symmedian point.

453. Another important circle, *the circle of Lemoine,* is defined by the following theorem:

Theorem.   *Let lines be drawn through the symmedian point, parallel to the sides of the triangle.   They meet the adjacent sides in six points lying on a circle whose center Z is the mid-point of KO.*

Let the line parallel to a_1 cut a_2 at P_3, and a_3 at Q_2; etc. First, P_1Q_1 is antiparallel to a_1, since $A_1Q_1KP_1$ is a parallelo-

gram, and A_1K bisects P_1Q_1; but a line bisected by a symmedian is antiparallel to the opposite side.   Hence each of the lines P_1Q_1, P_2Q_2, P_3Q_3 is equal to the radius of the cosine circle.   Now let F_1 be the mid-point of P_1Q_1, and Z that of OK.   Then F_1Z is parallel to A_1O and equal to half of it;

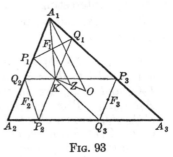

Fig. 93

it is therefore, like A_1O, perpendicular to the antiparallel line P_1Q_1.   If r denotes the radius of the cosine circle,

$$\overline{ZQ_1}^2 = \overline{F_1Z}^2 + \overline{F_1Q_1}^2 = \tfrac{1}{4}R^2 + \tfrac{1}{4}r^2,$$

which is the same for all six points P_1, Q_1, P_2, Q_2, P_3, Q_3. Hence all are on a circle about Z, with radius

$$\tfrac{1}{2} \sqrt{R^2 + r^2} = \frac{R \sec \omega}{2}$$

454. *The Lemoine circle divides any side into segments proportional to the squares of the sides:*

$$\overline{A_2 P_2} : \overline{P_2 Q_3} : \overline{Q_3 A_3} = a_3{}^2 : a_1{}^2 : a_2{}^2 \tag{344}$$

The chords cut from the sides by the Lemoine circle are proportional to the cubes of the sides.

The last result is obtained by treating the previous proportion by composition. Because of this relation, the circle is called by English writers the *triplicate-ratio* circle.

455. *The triangles $P_1 P_2 P_3$ and $Q_1 Q_2 Q_3$ are congruent, and similar to the given triangle in the ratio $1 : (2 \cos \omega)$; the respective centers of similitude are Ω, Ω'; the angle of similitude is ω. The length of each of the equal antiparallels $P_1 Q_1$, $P_2 Q_2$, $P_3 Q_3$ is $R \tan \omega$.*

456. The two Lemoine circles are representative and especially interesting members of the system known as *Tucker circles*; and we now discuss these circles from several points of view.

Theorem. *Let three equal lines $P_1 Q_1$, $P_2 Q_2$, $P_3 Q_3$ be drawn antiparallel to the sides of a triangle, so that any two, say $P_2 Q_2$ and $P_3 Q_3$, are on the same side of the third line, as $A_2 P_2 Q_3 A_3$. Then $P_2 Q_3 P_3 Q_2$ is an isosceles trapezoid; $P_3 Q_2$, $P_1 Q_3$, $P_2 Q_1$ are parallel to the respective sides. The mid-points C_1, C_2, C_3 of the antiparallels are on the respective symmedians and divide them proportionally; and if T divides KO in the same ratio, TC_1, TC_2, TC_3 are parallel to the radii OA_1, OA_2, OA_3 and equal. Since the antiparallels are perpendicular to the symmedians, they are equal chords of a circle with center T, which passes through the six given points. This circle is a Tucker circle.*

We see at once that the lines P_2Q_2 and P_3Q_3 make with A_2A_3 angles equal to α_1, and are equal; hence they are sides of an isosceles trapezoid. We know also that a symmedian bisects any antiparallel, and therefore C_1, C_2, C_3 are on the symmedians. But C_2C_3 is parallel to P_3Q_2 and to A_2A_3, and therefore divides A_2K and A_3K proportionally. If

$$\frac{\overline{KC_1}}{\overline{KA_1}} = \frac{\overline{KC_2}}{\overline{KA_2}} = \frac{\overline{KC_3}}{\overline{KA_3}} = \frac{\overline{KT}}{\overline{KO}} = c$$

the right triangles C_1P_1T, C_1Q_1T, etc., are congruent, with

$$\overline{TC_1} = c\,R, \quad \overline{P_1C_1} = \overline{C_1Q_1} = (1-c)\,R \tan \omega \qquad (450)$$

Hence *the radius of the Tucker circle is* $R\sqrt{c^2 + (1-c)^2 \tan^2 \omega}$. For any value of c, positive or negative, there is a Tucker circle.

457. Other methods of describing the Tucker circles are suggested by the various properties. Namely:

Theorem. *Starting with any point on a side of a triangle, we may describe a closed hexagon whose sides are alternately parallel and antiparallel to the sides. The antiparallels will be equal, and the six points lie on a Tucker circle. Or, if lines are drawn parallel to the sides of a triangle and distant from them proportionally to the lengths of the sides (in the proper sense), their extremities are points of a Tucker circle.*

Theorem. *Triangle $P_1P_2P_3$ is similar to $A_1A_2A_3$, with Ω as center of similitude; the angle of similitude θ between corresponding lines, and the ratio of similitude, are given by*

$$\tan \theta = \frac{1-c}{c} \tan \omega, \quad q = \frac{\overline{OT}}{\overline{O\Omega}} = \frac{\sin \omega}{\sin (\omega + \theta)}$$

For $\qquad \measuredangle\, P_2P_1P_3 = \measuredangle\, P_2Q_1P_3 = \measuredangle\, A_2A_1A_3$

so that the triangles are similar. By **33** the center of similitude is the point of intersection of the circles $A_1P_1P_3$, $A_2P_2P_1$,

$A_3P_3P_2$, which is a fixed point, the Miquel point. But in one case, that of the cosine circle, we know that this point is at Ω. The other relations are easily worked out. Similarly:

> *Triangle $Q_1Q_2Q_3$ is similar to $A_1A_2A_3$, with Ω' as center of similitude; the ratio of similitude is the same, and the angle of similitude is $-\theta$. Triangles $P_1P_2P_3$ and $Q_1Q_2Q_3$ are congruent.*

458. Theorem. *If the three lines ΩA_1, ΩA_2, ΩA_3 rotate as a rigid system about Ω, and intersect the sides A_1A_2, A_2A_3, A_3A_1 respectively at P_1, P_2, P_3; while $\Omega'A_1$, $\Omega'A_2$, $\Omega'A_3$ rotate through the same angle θ about Ω' in the opposite direction, meeting A_1A_3, A_3A_2, A_2A_1 respectively at Q_1, Q_2, Q_3; then triangles $P_1P_2P_3$ and $Q_1Q_2Q_3$ are congruent, and similar to the given triangle in the ratio shown above; the vertices of the triangles lie on a circle whose center lies on OK, and this circle has the properties given for Tucker circles. The locus of the negative Brocard point of $P_1P_2P_3$ is $\Omega'K$, that of the positive Brocard point of $Q_1Q_2Q_3$ is ΩK, while the locus of the symmedian point of either triangle is perpendicular to OK at K.*
We notice that the common pedal circle of Ω and Ω' is a Tucker circle, as are also the Lemoine circles, and the circumcircle of $A_1A_2A_3$.

Evidently the common pedal circle is the smallest of the Tucker circles, and all others are equal in pairs.

459. Coaxaloid circles. In the similar figures $A_1A_2A_3O\Omega$ and $P_1P_2P_3T\Omega$, homologous lines $O\Omega$ and $T\Omega$ are proportional to the radii of the circles. Hence if r is the radius of that one of the Tucker circles whose center is T,

$$r = \frac{R}{\overline{O\Omega}}\,\overline{T\Omega}$$

that is, r bears a constant ratio to $T\Omega$. If r were *equal* to $\overline{T\Omega}$, we should have a set of coaxal circles through the points Ω, Ω'. Thus:

Theorem. *The Tucker circles of a triangle are derived from coaxal circles through the common points Ω and Ω' by increasing the radius of each by the constant multiplier, $R/\overline{O\Omega}$.* If the circles of a coaxal system are changed by multiplying each radius by a constant, the resulting system of circles is called a *coaxaloid system*. Its properties are treated at length by Third (**501**). The Tucker circles are a representative coaxaloid system.

460. The *Taylor circle* is another interesting member of the system of Tucker circles.

Theorem. *If from the foot of each altitude lines are drawn perpendicular to the adjacent sides, their feet lie on a circle, which is a Tucker circle.*

For if H_1P_1 and H_1Q_1 are perpendicular to A_1A_2 and A_1A_3 respectively, the figures $A_1H_3HH_2$ and $A_1P_1H_1Q_1$ are similar; P_1Q_1 is parallel to H_2H_3, and we easily find that

$$\overline{P_1Q_1} = 2\,R \sin\,\alpha_1 \sin\,\alpha_2 \sin\,\alpha_3$$

so that the three antiparallels are equal.

Theorem. *The line $\overline{P_1Q_1}$ bisects $\overline{H_1H_2}$ and $\overline{H_1H_3}$, and in an acute triangle*

$$\overline{P_1Q_1} = \tfrac{1}{2}(\overline{H_2H_3} + \overline{H_3H_1} + \overline{H_1H_2}) \quad \text{(Feuerbach)}.$$

The center of the Taylor circle is the incenter or an excenter of the triangle whose vertices are the mid-points of H_2H_3, H_3H_1, H_1H_2; that is, it is the center of the Spieker circle of $H_1H_2H_3$; and the lines connecting these mid-points with the center of the circle are perpendicular to the sides of the given triangle.

THE BROCARD TRIANGLES AND BROCARD CIRCLE

461. In this section we establish the existence of two notable triangles, both inscribed in the Brocard circle.

We recall that $A_2\Omega$ and $A_3\Omega'$ make the angle ω with A_2A_3;

let their point of intersection be B_1, and let B_2 and B_3 be similarly found. Then $B_1B_2B_3$ is the *First Brocard Triangle.*

Theorem. *The triangles $B_1A_2A_3$, $B_2A_3A_1$, $B_3A_1A_2$ are isosceles and similar with base angles ω. The sum of their areas is Δ.*

$$\overline{B_1O_1} = \tfrac{1}{2}\, a_1 \tan \omega = \overline{KK_1} \tag{438}.$$

Hence KB_1 is parallel to A_2A_3. The triangle $B_1B_2B_3$ is inscribed in the Brocard circle drawn on OK as diameter. Triangles $A_1A_2A_3$ and $B_1B_2B_3$ are inversely similar.

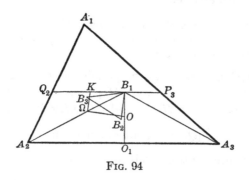

FIG. 94

We see that KB_1O is a right angle; and therefore B_1 is on the circle on OK as diameter, as are also B_2 and B_3. Moreover,

$$\measuredangle\ B_2B_1B_3 = \measuredangle\ B_2KB_3 = \measuredangle\ B_2K, B_3K = \measuredangle\ A_3A_1A_2$$

462. *The first Brocard triangle is in perspective with the given triangle, A_1B_1, A_2B_2, A_3B_3 being concurrent at a point D.*

This may be proved in several ways; for instance, as a case of **357.** Again, the triangles are twice in perspective at Ω and Ω', and therefore (**381**) triply in perspective. Finally, the line KB_1 passes through the points P_3 and Q_2 of the Lemoine circle, and the Lemoine and Brocard circles are concentric. Hence A_1K and A_1B_1 are isotomic, and the point D is the isotomic conjugate of K.

463. The Brocard points are determined by two triads of circles which we designated (**433**) as c_1, etc.   Let the two circles c_1, c_1' tangent at A_1 to A_1A_2 and A_1A_3, and passing respectively through A_3, A_2, meet again at C_1; then the triangle $C_1C_2C_3$ is the *Second Brocard Triangle*.

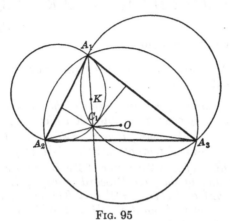

FIG. 95

Theorem. $\angle A_1C_1A_2 = \angle A_1C_1A_3 = 180 - \alpha_1$

   $\angle A_2C_1A_3 = 2\,\alpha_1.$

That is $\measuredangle A_2C_1A_1 = \measuredangle A_1C_1A_3 = \measuredangle A_2A_1A_3,$

   $\measuredangle A_2C_1A_3 = 2 \measuredangle A_2A_1A_3.$

Hence the point C_1 lies on the circle A_2A_3O.   Triangles $A_1A_2C_1$ and $A_3A_1C_1$ are directly similar in the ratio a_3/a_2.   The perpendiculars from C_1 to A_1A_2 and A_1A_3 are proportional to those sides (being homologous altitudes), *and C_1 lies on the symmedian A_1K. OC_1 is perpendicular to A_1K (by means of the equation $\measuredangle OC_1A_1 = \measuredangle OC_1A_2 + \measuredangle A_2C_1A_1$). C_1 is the mid-point of the chord of the circumcircle from A_1 through K.   Each vertex of the second Brocard triangle lies on the Brocard circle.   (For OC_1 is perpendicular to C_1K).*

464. Theorem.   *The median point of the first Brocard triangle is at M (**358**).*

465. Theorem.   *The Brocard triangles are in perspective at M.*

For M is the common median point of the inversely similar triangles $A_1A_2A_3$ and $B_1B_2B_3$;

$$\angle B_3B_1M = \angle MA_1A_3 = \angle A_2A_1K = \angle B_3KC_1$$
$$= \angle B_3B_1C_1$$

so that B_1, C_1, M are collinear.

466. Further properties of this figure may be worked out in abundant detail: *

The lines connecting the mid-points of the sides of $B_1B_2B_3$ with those of $A_1A_2A_3$ meet at a point R which is on DM (462), $\overline{MR} = \frac{1}{3}\overline{DM}$. Moreover, R is the mid-point of $\Omega\Omega'$. M is the median point of triangle $\Omega\Omega'D$. DH is parallel to OK, $\overline{DH} = 2\,\overline{OR}$. If Z is the center of the Brocard circle, and H' the orthocenter of the first Brocard triangle, H' is the intersection of ZM and HD. HH' and KO are equal and parallel; the common mid-point of OH and KH' is the center F of the nine point circle.

467. Problem.   *To construct a triangle having a given triangle as its first Brocard triangle.*

The solution depends on the fact that any triangle is inversely similar to its first Brocard triangle. We locate the Brocard points of the given triangle, also its first Brocard triangle; then by similarity we locate on the circumcircle of the given triangle the Brocard points of the required triangle. The vertices of the latter can then be found at once.

468. Theorem.   *In a scalene triangle, the symmedian point lies on the arc of the Brocard circle which is between the largest and the smallest of the angles of the first Brocard triangle.*

Suppose $\alpha_1 < \alpha_2 < \alpha_3$, and let A_2A_3 be horizontal, with A_2

---

* Cf. Fuhrmann or Emmerich. The reader is not expected to prove these results.

at the left. Then the whole symmedian lies to the left of OO_1 (**344**) and thus KB_1 is drawn toward the right. Hence KB_1 is in the direction A_2A_3, KB_2 in the direction A_1A_3, and KB_3 in the direction A_2A_3. Hence

$$\angle B_1KB_2 = \alpha_3, \ \angle B_2KB_3 = \alpha_1,$$
$$\angle B_1KB_3 = 180° - \alpha_2 = \alpha_1 + \alpha_3$$

and K is opposite to B_2.

THE STEINER POINT AND THE TARRY POINT

469. Theorem. *If lines are drawn through the vertices of a triangle parallel to the corresponding sides of the first Brocard triangle, they meet at a point on the circumcircle (Steiner point *).*

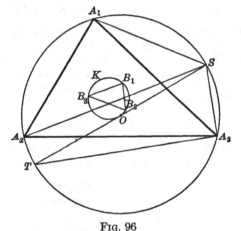

Fig. 96

We may at the same time prove the more general theorem:

Lines through the vertices of a given triangle, and parallel to the corresponding sides of a triangle inversely similar to it, are concurrent on its circumcircle.

* Steiner (*Collected Works*, 2, p. 689) considered this point from a different aspect.

For if $A_1A_2A_3$ and $B_1B_2B_3$ are inversely similar, and A_1S and A_2S are parallel respectively to B_2B_3 and B_3B_1, then

$$\measuredangle\, A_1SA_2 = \measuredangle\, B_2B_3B_1 = \measuredangle\, A_1A_3A_2$$

S lies on the circumcircle of $A_1A_2A_3$, and A_3S is parallel to B_1B_2.

470. Theorem.　*The symmedian point K is the Steiner point of the first Brocard triangle.*

471. Theorem.　*The lines through the vertices of a triangle, perpendicular to the corresponding sides of the first Brocard triangle, are concurrent at a point called the Tarry point, T, diametrically opposite to the Steiner point on the circumcircle. The circumcenter is the Tarry point of $B_1B_2B_3$.*

472. Theorem.　*The Simson lines of S and T are respectively parallel and perpendicular to OK.*　**(326)**

Additional properties.　*The Tarry point lies on the line ZMH'* **(466)**, *the Euler line of triangle $B_1B_2B_3$. The diameter ST passes through the point D; the isogonal conjugate D' of D lies on OK, and is harmonic to R, so that the tangents to the Brocard circle at Ω and Ω' meet at D'. T, H, D' are collinear.*

SOME RELATED TRIANGLES

473. We now consider briefly certain triangles associated with a given triangle, which have the same Brocard angle.

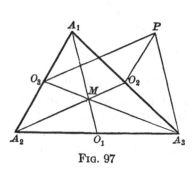

FIG. 97

Theorem.　*A triangle can be constructed whose sides are equal and parallel to the medians of a given triangle.*

If O_3P is equal and parallel to A_2O_2 and in the same direction, then PA_3O_3 is such a triangle. It is called the median triangle of $A_1A_2A_3$, and its sides are given by **96** a,

$$m_1 = \tfrac{1}{2} \sqrt{2\, a_2^2 + 2\, a_3^2 - a_1^2}, \text{ etc.}$$

Corollary. *The median triangle of the median triangle is similar to the given triangle, in the ratio 3/4. The area of the median triangle PA_3O_3 is $3/4\ \Delta$.*

474. Theorem. *The median triangle has the same Brocard angle as the given triangle.*

For if w is the Brocard angle of the median triangle,

$$\cot w = \frac{m_1^2 + m_2^2 + m_3^2}{4\,\Delta'} = \frac{\tfrac{1}{4}(3\,a_1^2 + 3\,a_2^2 + 3\,a_3^2)}{4 \cdot \tfrac{3}{4}\Delta} \quad (435)$$

Exercise. *Show that the only triangle similar to its median triangle in the sense $A_1A_2A_3 \sim M_1M_2M_3$ is equilateral.*

475. We have already proved (**351**) that if the symmedians of a triangle are extended to meet the circumcircle at P_1, P_2, P_3, then K is also the symmedian point of $P_1P_2P_3$. Therefore $P_1P_2P_3$ and $A_1A_2A_3$, known as *cosymmedian triangles*, have the same Brocard circle, and obviously the same second Brocard triangle. Moreover, from **450** we may see that the Brocard angle is known when R and $\overline{OK}$ are known, which establishes the result that $P_1P_2P_3$ and $A_1A_2A_3$ have the same Brocard angle, and therefore that their Brocard points coincide.

Theorem. *Either of two cosymmedian triangles is similar to the median triangle of the other.*

For we know (**199**) that $P_1P_2P_3$ is similar to the pedal triangle $K_1K_2K_3$ of K; but K is the median point of $K_1K_2K_3$ (**350**), hence the medians of $P_1P_2P_3$ are proportional to KK_1, KK_2, KK_3, which in turn (**342**) are proportional to a_1, a_2, a_3.

Corollary. *Cosymmedian triangles have the same circumcircle, and a common symmedian point as center of perspective; they have the same Brocard points, Brocard angle, and second Brocard triangle.*

476. Theorem. *If points divide the sides of a given triangle in equal ratios, they are vertices of a triangle having the same Brocard angle as the given triangle.*

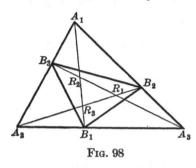

FIG. 98

If the sides are divided at B_1, B_2, B_3 in the ratio $-m/n$, where we take $m + n = 1$, then $A_2B_1 = ma_1$, $B_1A_3 = na_1$, etc.; it is to be proved that $B_1B_2B_3$ has the Brocard angle ω. The proof is long but straightforward. Expressing the lengths B_2B_3, etc., by means of the law of cosines, we find the Brocard angle of $B_1B_2B_3$ by the formula

$$\cot \omega' = \frac{\overline{B_2B_3}^2 + \overline{B_3B_1}^2 + \overline{B_1B_2}^2}{4 \text{ area } B_1B_2B_3}$$

which after some reduction reduces to

$$\frac{(a_1^2 + a_2^2 + a_3^2)(1 - 3mn)}{4\,\Delta\,(1 - 3mn)}$$

or cot ω (**435**).

Exercises. *In the foregoing figure, let A_2B_2 meet A_3B_3 at R_1, etc.; then triangle $R_1R_2R_3$ has the Brocard angle ω, as does also the triangle whose sides are A_1B_1, A_2B_2, A_3B_3.*

The method consists in expressing the ratios of various lines and areas in terms of m and n, and eventually showing that A_1, A_2, A_3 divide the sides of triangle $R_1R_2R_3$ in equal ratios.

477. Let us recall the two triads of c-circles which touch the sides of the triangle at the vertices and pass through the Brocard points. Let the circles c_2 and c_3' which are tangent to A_2A_3 at A_2 and A_3 respectively, and pass through A_1, meet again at D_1; etc. Then the vertices $A_1A_2A_3$, the Brocard points, the second Brocard triangle, and the D-triangle, account for all the intersections of the c-circles. We summarize the properties of the triangle $D_1D_2D_3$, which can be demonstrated without difficulty.

Theorem. $\measuredangle A_2D_1A_3 = \measuredangle A_3A_1A_2$, and D_1 lies on the circle through A_2, A_3, H. D_1 is on the median A_1O_1, and is the foot of the perpendicular from H on the median.* The circle drawn on HM as diameter passes through D_1, D_2, D_3; and triangle $D_1D_2D_3$ is inversely similar to the median triangle. D_1 is the reflection with regard to A_2A_3 of the point where the symmedian A_1K cuts the circumcircle. Corresponding vertices of $D_1D_2D_3$ and of the second Brocard triangle are isogonal conjugates. The pedal triangles of D_1, D_2, D_3 are isosceles; that of D_1 has base angles equal to α_1 or its supplement.

478. The centers of the c-circles are vertices of two interesting triangles. Let U_1, U_2, U_3 be the centers of c_1, c_2, c_3; and V_1, V_2, V_3 of c_1', c_2', c_3'. Thus U_1 is the intersection of the perpendicular bisector of A_1A_3 with the perpendicular to A_1A_2 at A_1; and so on.

Theorem. *Triangles $U_1U_2U_3$ and $A_1A_2A_3$ are similar, with Ω as center of similitude. The negative Brocard point of $U_1U_2U_3$ is at O. $V_1V_2V_3$ is similar to $A_1A_2A_3$, and its Brocard points are O and Ω'. $U_1U_2U_3$ and $V_1V_2V_3$ are congruent, and their center of similitude is the center of the Brocard circle. The centers of circles through $U_1U_2U_3$ and $V_1V_2V_3$ respectively lie equidistant from O on a line parallel to $\Omega\Omega'$. The center of the Brocard circle is their common symmedian point. Triangles $U_3U_1U_2$ and $V_2V_3V_1$ are in perspective at O, hence corresponding lines meet at concurrent points; these points are on OK. (For U_1U_2 is the perpendicular bisector of $A_1\Omega$, and V_1V_3 that of $A_1\Omega'$; their intersection therefore is equidistant from Ω and Ω'.) The sides of the two triangles meet the Lemoine circle where it cuts the sides of the given triangle.*

* Note the analogy; C_1 is the foot of the perpendicular from O on the symmedian (**463**).

Exercise.   The reader of this and the following two chapters will recognize that no sharp distinction is made between proved theorems and those whose proofs are left as exercises. However, opportunity for original work is to be found in the following sections, of which those in parenthesis are somewhat more difficult: **(435)**, **436**, **437–444**, **447**, **449**, **450**, **454**, **455**, **457**, **458**, **460**, **462–464**, **(466)**, **467**, **470**, **471**, **(472)**, **475–478**.

CHAPTER XVII
EQUIBROCARDAL TRIANGLES

479. In this chapter we consider systems of triangles having a common Brocard angle. First, the triangles on a given base with a given Brocard angle give rise to the Neuberg circles. Then we obtain all the triangles in a plane having the same Brocard angle, by a simple projection in space. A consideration of the circles of Apollonius leads to the study of the Brocard angle of the pedal triangle of a point; and thus to the solution of the problem of the locus of a point whose pedal triangle has a given Brocard angle.

THE NEUBERG CIRCLES

480. Theorem. *The locus of the vertex A_1 of a triangle on a given base A_2A_3 and with a given Brocard angle ω is a circle on either side of A_2A_3, from whose center N_1 the base A_2A_3 subtends the angle $2\,\omega$; the radius of the circle is*

$$v = \frac{a_1}{2}\sqrt{\cot^2 \omega - 3}.$$

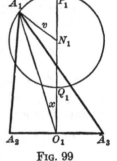

FIG. 99

For let $A_1A_2A_3$ be a triangle with Brocard angle ω. We denote the angle A_1O_1O by x and, as before, A_1O_1 by m_1; then

$$h_1 = m_1 \cos x, \quad 2\,\Delta = a_1 m_1 \cos x.$$

Now $\quad a_1^2 + a_2^2 + a_3^2 = 4\,\Delta \cot \omega \quad$ **(435)**

also $\quad a_2^2 + a_3^2 = \tfrac{1}{2}\,a_1^2 + 2\,m_1^2 \quad$ **(96)**

whence $\quad \tfrac{3}{2}\,a_1^2 + 2\,m_1^2 = 2\,a_1 m_1 \cos x \cot \omega$

We now throw this equation into the form of the law of cosines **(15** c**)** introducing $u = \overline{O_1 N_1} = \dfrac{a_1}{2} \cot \omega$, and v as given above. The equation may be rewritten

$$m_1{}^2 - 2m\left(\frac{a_1}{2}\cot\omega\right)\cos x + \frac{a_1{}^2}{4}\cot^2\omega = \frac{a_1{}^2}{4}(\cot^2\omega - 3)$$

or $\qquad m_1{}^2 + u^2 - 2m_1u\cos x = v^2$

showing that $\overline{A_1N_1}$ equals v, which is constant when a_1 and ω are given.

The three circles determined by this theorem, each passing through a vertex of a given triangle, are called the *Neuberg circles* n_1, n_2, n_3 of the triangle. Each is the locus of the vertex of a triangle on the opposite side and having the given Brocard angle. Some of their properties are immediately obvious.

Theorems. *The power of A_2 or A_3 with regard to the circle n_1 is $\overline{A_2A_3}^2$ (being $\overline{A_2N_1}^2 - v^2$). The median point of triangle $N_1N_2N_3$ is at M* (**358**), *and A_1N_1, A_2N_2, A_3N_3 are concurrent* (**357**).

(Actually the point of concurrence is the Tarry point, so that the tangents to the Neuberg circles at the respective vertices are concurrent at the Steiner point.)

481. Theorem. *If the value of the Brocard angle ω is given, the largest possible value, δ, and the smallest, δ', of any angle of the triangle are given by*

$$\cot\frac{\delta}{2} = \cot\omega - \sqrt{\cot^2\omega - 3}, \quad \cot\frac{\delta'}{2} = \cot\omega + \sqrt{\cot^2\omega - 3}$$

For if O_1ON_1 cuts the Neuberg circle at Q_1 and P_1, then $A_2Q_1A_3 = \delta$, and $A_2P_1A_3 = \delta'$. By trigonometric methods we have further:

$$\cot\delta = \tfrac{1}{3}(\cot\omega - 2\sqrt{\cot^2\omega - 3}),$$

$$\cot\delta' = \tfrac{1}{3}(\cot\omega + 2\sqrt{\cot^2\omega - 3}),$$

$$\sin\delta\,\sin\delta' = 3\sin^2\omega,$$

$$\cos\delta\,\cos\delta' = 5\sin^2\omega - 1.$$

For example, if $\cot\omega = 1.75$, $\omega = 29°\,44'\,42''$, the angles of

the triangle are between limits which are approximately 53° 7'
and 67° 23'.   In this case $\overline{OK} = \tfrac{4}{7}R$; $\overline{\Omega\Omega'} = \tfrac{8}{65}R$.

482. Theorem.   *If an angle α of a triangle is given, the
maximum possible Brocard angle is given by*

$$\cot \omega = \frac{3}{2}\,\tan\frac{\alpha}{2} + \frac{1}{2}\,\cot\frac{\alpha}{2}$$

483. Theorem.   *On one side of a given line as base it is possible to construct six triangles directly or inversely similar to
a given scalene triangle; their vertices lie on their common
Neuberg circle.*

Let triangles $A_1A_2A_3$, $A_2A_3B_1$, $A_3C_1A_2$ be directly similar,
and let $D_1A_3A_2$, $A_3A_2E_1$, $A_2F_1A_3$ be inversely similar to them.
Then since the six triangles have the same
Brocard angle, A_1, B_1, C_1,D_1, E_1, F_1 lie on
the Neuberg circle. Also A_1E_1, B_1F_1, C_1D_1
pass through A_2, and A_1F_1, B_1D_1, C_1E_1
through A_3; that is, the sides of the hexagon
$A_1E_1C_1D_1B_1F_1$ pass alternately through
the fixed points A_2 and A_3. Triangles
$A_1C_1B_1$ and $D_1F_1E_1$ are similar to $A_1A_2A_3$.
If A_2A_1 and A_3D_1 meet at X, A_2B_1 and A_3E_1
at Y, and A_2C_1 and A_3F_1 at Z, triangles
A_2A_3X, A_2A_3Y, A_3A_3Z are isosceles,
with base angles α_2, α_1, α_3 respectively.

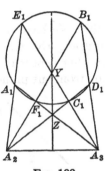

Fig. 100

An exceptional case arises when the triangles are isosceles.
If $A_2A_3P_1$ and $A_2A_3Q_1$ are isosceles triangles, so that P_1 and Q_1
are on O_1N_1, and if A_2P_1 meets the circle at R_1, A_3R_1 is tangent
to the circle and $A_2A_3R_1$ is isosceles, $A_2A_3 = A_3R_1$.

484. Theorem.   *The Neuberg circle n_1 is orthogonal to the
circles with centers A_2 and A_3, radii equal to A_2A_3.   Hence
the Neuberg circles on a given base, for different values of ω,
are a coaxal system each of whose limiting points L, L' forms
with A_2A_3 an equilateral triangle.*

485. Theorem. *If the vertex A_1 of a triangle describes the Neuberg circle n_1, its median point describes a circle whose radius is one-third that of the Neuberg circle. Such a circle is called a McCay circle; the three McCay circles of a given triangle are concurrent at the median point M.*

These circles will be treated in the next chapter.

Exercises. *The angle between O_1O and a tangent to the Neuberg circle n_1 from O_1 is given by $\cos \phi = \sqrt{3} \tan \omega$.*

The distances from the circumcenter to the Neuberg centers are proportional to the cubes of the sides: $\overline{N_1O} = a_1{}^3/4\,\Delta.$

$$\text{Hence} \qquad \cot \omega = \frac{\overline{N_1O}}{a_1} + \frac{\overline{N_2O}}{a_2} + \frac{\overline{N_3O}}{a_3};$$

$$\overline{N_1O} \cdot \overline{N_2O} \cdot \overline{N_3O} = R^3$$

VERTICAL PROJECTIONS

486. We have seen in the foregoing section one method of grouping the triangles that have the same Brocard angle. Another aspect of the problem is based on a parallel projection from one plane to another. We find that if all equilateral triangles in a plane are projected into a second plane by lines perpendicular to the latter, the resulting triangles will have equal Brocard angles.

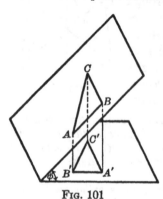

For convenience of expression we shall speak of a horizontal plane and an oblique plane making an angle ϕ with it. By a *projection* we shall understand a vertical projection by means of rays perpendicular to the horizontal plane.

FIG. 101

Obviously a straight line of either plane projects into a straight line in the other plane, and a circle into an oval curve (ellipse).

If a line of length a in the oblique plane is parallel to the line of intersection l of the two planes, its projection in the horizontal plane is parallel to l and equal to a; if perpendicular, the projection is also perpendicular to l, and its length is $a \cos \phi$. The formulas for the direction and length of an oblique line are easily written.

If a triangle in the oblique plane has area D, the area of its projection is $D \cos \phi$; by combination and the method of limits, we obtain the same formula for areas bounded by polygons and curves.

If two figures in either plane are similar, the corresponding figures will generally not be similar, except when the given figures are homothetic. On two corresponding lines, however, corresponding segments are proportional.

487. Theorem. *If two equilateral triangles of the oblique plane, described in the same direction, are projected into the horizontal, the resulting tri-angles have the same Bro-card angle.*

For let $A_1A_2A_3$ and $B_1B_2B_3$ be the equilateral triangles, M and N their centers. If lines through M, parallel respectively to NB_1, NB_2, NB_3, meet A_2A_3, A_3A_1, A_1A_2 at C_1, C_2, C_3, triangles $B_1B_2B_3$ and $C_1C_2C_3$ are homothetic. Obviously C_1, C_2, C_3 divide A_2A_3, A_3A_1, A_1A_2 in equal ratios. Let the projections of all these points in the horizontal plane be denoted by primes. Then $B_1'B_2'B_3'$ and $C_1'C_2'C_3'$ are homothetic;

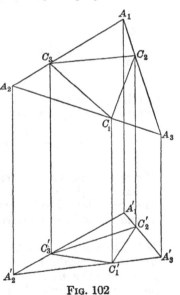

Fig. 102

C_1', C_2', C_3' divide $A_2'A_3'$, $A_3'A_1'$, $A_1'A_2'$ in equal ratios, and therefore (**476**) $C_1'C_2'C_3'$ and $A_1'A_2'A_3'$ have the same Brocard angle. Therefore $A_1'A_2'A_3'$ and $B_1'B_2'B_3'$ have the same Brocard angle ω.

Corollary. *The Brocard angle ω depends only on the angle ϕ between the planes, viz.:*

$$\cot \omega = \frac{\sqrt{3}}{2}\left(\cos\phi + \frac{1}{\cos\phi}\right)$$

For consider, in particular, an equilateral triangle having one side a parallel to l. It projects into an isosceles triangle whose base is a, and whose altitude is $\frac{a}{2}\sqrt{3}\cos\phi$. Then its equal sides are given by

$$b^2 = \frac{a^2}{4} + \frac{3a^2}{4}\cos^2\phi$$

whence

$$\cot\omega = \frac{a^2 + b^2 + b^2}{4\,\Delta'} = \frac{\frac{3}{2}a^2(1 + \cos^2\phi)}{\sqrt{3}\,a^2\cos\phi} = \frac{\sqrt{3}}{2}\frac{1 + \cos^2\phi}{\cos\phi}$$

Corollary. *When ω is given, we find a single value for $\cos\phi$ between 1 and 0, namely:*

$$\cos\phi = \frac{1}{\sqrt{3}}(\cot\omega - \sqrt{\cot^2\omega - 3})$$

Hence it is possible, by proper choice of ϕ, to project equilateral triangles into triangles having as Brocard angle any given angle less than 30°; there is for any ω only one value of ϕ, and it can be constructed with ruler and compass.

488. Theorem. *Any triangle in a horizontal plane can be projected vertically into an equilateral triangle in an oblique plane; in other words, any triangular prism can be cut by a plane in an equilateral triangle.*

For let ω be the Brocard angle of the given triangle, Δ its area; and denote by d the side of the required equilateral triangle. We find

$$d^2 = \frac{4\,\Delta}{3}\,(\cot\omega - \sqrt{\cot^2\omega - 3})$$

Taking as center the vertex A_1 of the largest angle of the given triangle, let a sphere of radius d cut the other two edges of the prism at points P_2 and P_3 on the same side of the horizontal plane. Then $A_1P_2P_3$ is equilateral, as can be verified by direct computation.

Our result may then be stated as follows: *All equilateral triangles in the oblique planes that make a fixed angle with the horizontal plane, project into the latter plane in triangles that have a constant Brocard angle; and conversely, all triangles with that Brocard angle can be so derived. If a single equilateral triangle in the oblique plane is rotated about any point, the projected triangle takes on all possible forms of triangles with the specified Brocard angle.*

489. Theorem. *If the vertices of one triangle lie on the sides of another, and divide them in equal ratios (476) they may be projected simultaneously into equilateral triangles. Conversely, if two triangles described in the same direction have the same Brocard angle, a triangle directly similar to either can be inscribed in the other, dividing its sides in equal ratios.*

For we may project the given triangles into equilateral triangles, in oblique planes which will not generally be parallel, but will make with the horizontal the same angle ϕ; then we revolve one of them about a vertical axis until they are parallel, the original triangle moving congruent to itself in the horizontal plane. We may then inscribe in one equilateral triangle another which is homothetic to the second, and divides proportionally the sides of the first. The same situation holds in the non-similar triangles in the horizontal plane.

Corollary.　*The triangles whose vertices divide in equal ratios the sides of a given triangle constitute all the different forms of triangle having the same Brocard angle.*

THE CIRCLES OF APOLLONIUS AND ISODYNAMIC POINTS *

490. We have already seen (**59**) that the locus of a point P whose distances from the vertices A_2 and A_3 of a triangle are proportional to A_1A_2 and A_1A_3, is a circle through A_1, which has as a diameter on A_2A_3 the line between the extremities X_1 and Y_1 of the bisectors of angle A_1. In a triangle there are three such circles, called the circles of Apollonius. We shall designate them by k_1, k_2, k_3; their centers, on the respective sides of the triangle, by L_1, L_2, L_3.

Theorem.　*The center L_1 is the intersection of the side A_2A_3 with the tangent to the circumcircle at A_1; the circles of Apollonius are orthogonal to the circumcircle.*

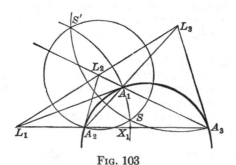

FIG. 103

For $\angle L_1A_1X_1 = \angle A_1X_1A_2 = \alpha_3 + \dfrac{\alpha_1}{2}$, $\quad \angle L_1A_1A_2 = \alpha_3$

The center L_1 is the pole of the symmedian A_1K with regard to the circumcircle.

For if the tangents at A_2 and A_3 meet at T_1, A_1T_1 is the

* The theory of poles and polars (**134** ff.) is used in this section.

symmedian.   The polars of A_1 and T_1 are respectively A_1L_1 and A_2A_3; hence the pole of the line is L_1.

> *The centers L_1, L_2, L_3 are collinear on the line polar of K with regard to the circumcircle; this line is perpendicular to the Brocard axis OK, and is the radical axis of the circumcircle and the Brocard circle.   It is called the Lemoine line, and is also the trilinear polar of K with regard to $A_1A_2A_3$.*
> *The circles of Apollonius are coaxal; their radical axis is the Brocard line OK, and they intersect at two points of this line which are inverse with regard to the circumcircle.*

491. Theorem.   *The circle of Apollonius k_1 is the locus of a point whose pedal triangle is isosceles in the sense $\overline{P_1P_2} = \overline{P_1P_3}$.*

For $$\overline{P_1P_2} = \overline{PA_3} \sin \alpha_3 \qquad (190)$$

and $\overline{P_1P_2}$ and $\overline{P_1P_3}$ are equal if and only if

$$\frac{\overline{PA_2}}{\overline{PA_3}} = \frac{\sin \alpha_3}{\sin \alpha_2} = \frac{\overline{A_1A_2}}{\overline{A_1A_3}}$$

Theorem.   *More generally, if two points P, Q are inverse with regard to the circle k_1, their pedal triangles are inversely similar in the sense $P_1P_2P_3 \sim Q_1Q_3Q_2$; and conversely.*

For A_2 and A_3 are inverse with regard to k_1; hence P, Q, A_2, A_3 are concyclic, and $\angle A_2PA_3 = \angle A_2QA_3$.

Hence $$\angle P_2P_1P_3 = \angle Q_2Q_1Q_3 \qquad (186)$$

Again, by **75**, $\angle A_1PA_2 + \angle A_1QA_3 = \angle A_1L_1A_3$, whence by **186** again, $\angle P_1P_3P_2 = \angle Q_3Q_2Q_1$; and similarly

$$\angle P_1P_2P_3 = \angle Q_2Q_3Q_1.$$

Conversely let the pedal triangles of P and Q be similar in this sense, and let R be the inverse of P with regard to the circle k_1; then Q and R have similar pedal triangles and therefore coincide.

492. Definition.   The *Isodynamic Points* of a triangle are the points S, S' common to the circles of Apollonius. They are inverse with regard to the circumcircle, on the Brocard line OK, and equidistant from the Lemoine line $L_1L_2L_3$.

Theorem. *The distances from either isodynamic point to the vertices are inversely proportional to the sides.*

Theorem. *The pedal triangle of either isodynamic point is equilateral.*

For S, $\measuredangle\, S_1S_2S_3 = 60°$; for S', $\measuredangle\, S_1'S_2'S_3' = 120°$.

This theorem follows from **491** or from **190.** Hence

$$\measuredangle\, A_2SA_3 = \measuredangle\, A_2A_1A_3 + 60°, \quad \measuredangle\, A_2S'A_3 = \measuredangle\, A_2A_1A_3 + 120°$$

If the lines from the vertices through an isodynamic point meet the circumcircle at X_1, X_2, X_3, then $X_1X_2X_3$ is equilateral.

Theorem. *An inversion with regard to either isodynamic point as center transforms the given triangle into an equilateral triangle.* **(200)**

493. Theorem. *The isodynamic points are the isogonal conjugates of the isogonic centers.* **(354 ff.)**

For if T is the isogonal conjugate of S,

$$\measuredangle\, A_2SA_3 + \measuredangle\, A_2TA_3 = \measuredangle\, A_2A_1A_3, \quad \measuredangle\, A_2TA_3 = 120°; \text{ etc.}$$

Theorem. *The vertices of the D-triangle* **(477)** *lie on the respective circles of Apollonius.*

494. Theorem. *The vertex C_1 of the second Brocard triangle is the inverse of O with regard to the circle k_1.*

For we saw **(463)** that C_1 is the mid-point of the symmedian chord of the circumcircle. But this line is the polar of O with regard to k_1.

Corollaries. *Hence the pedal triangles $X_1X_2X_3$, $Y_1Y_2Y_3$, $Z_1Z_2Z_3$ of C_1, C_2, C_3 are inversely similar to the given triangle, in the following senses:*

$$A_1 A_2 A_3 \sim X_1 X_3 X_2 \sim Y_3 Y_2 Y_1 \sim Z_2 Z_1 Z_3.$$

There are six points on the Brocard circle whose pedal triangles are similar to the original triangle, and are described in the same sense, namely O, Ω, Ω', C_1, C_2, C_3. For the first three the similarity is direct, for the others it is inverse. The triads of points are inverse with regard to the circles of Apollonius. Thus the triangles $O\Omega\Omega'$ and $C_1 C_2 C_3$ are triply in perspective; OC_1, ΩC_2, $\Omega' C_3$ pass through L_1, and so on. Therefore the triangles have three axes of perspective, which are concurrent at a point P of OK;

$$\overline{OP} = \frac{\overline{OK}}{1 + 3\,tan^2\,\omega}$$

The circles of Apollonius are orthogonal to the Brocard circle. The circumcircle, the Brocard circle, the Lemoine line, and the isodynamic points belong to a coaxal system, orthogonal to the circles of Apollonius.

We shall call these circles the *Schoute coaxal system.*

495. *The inverse of each of these six points with regard to the circumcircle also has the property that its pedal triangle is similar to the given triangle, but is inscribed in the opposite sense. Hence there are eleven points whose pedal triangles are similar to the given triangle, six on the circumcircle and five on the Lemoine line.*

496. The generalization of the foregoing result is fairly obvious.

Since any two points inverse with regard either to the circumcircle or to a circle of Apollonius have similar pedal triangles, any chosen point may be successively inverted with regard to these circles, to the following effect:

Theorem. *In general, there are twelve points whose pedal triangles with regard to a given triangle have a given form. They lie six by six on two circles of the Schoute system, which are mutually inverse with regard to the circumcircle; and the six points on either circle constitute two triangles triply in perspective, and mutually inverse with regard to each of the circles of Apollonius.*

THE CIRCLES OF SCHOUTE

497. We next inquire as to the locus of a point whose pedal triangle has a constant Brocard angle. It turns out, as suggested by the theorem just given, that this locus is a circle of the Schoute coaxal system.

Theorem. *If O is the center of an equilateral triangle $C_1C_2C_3$, the Brocard angle of the pedal triangle of any point P depends only on the value of $\overline{OP}$;*

$$\cot \omega = \sqrt{3}\,\frac{R^2 + \overline{OP}^2}{R^2 - \overline{OP}^2}$$

For $\overline{P_2P_3} = \overline{C_1P} \sin 60°$, hence

$$\overline{P_2P_3}^2 + \overline{P_3P_1}^2 + \overline{P_1P_2}^2 = \tfrac{3}{4}(\overline{C_1P}^2 + \overline{C_2P}^2 + \overline{C_3P}^2)$$
$$= \tfrac{3}{4}(\overline{OC_1}^2 + \overline{OC_2}^2 + \overline{OC_3}^2 + 3\,\overline{PO}^2) \qquad \textbf{(275)}$$
$$= \tfrac{9}{4}(R^2 + \overline{OP}^2)$$

Again, by **198,**

area $P_1P_2P_3 = \tfrac{1}{2}(R^2 - \overline{OP}^2) \sin^3 60° = \dfrac{3\sqrt{3}}{16}(R^2 - \overline{OP}^2)$

Hence the result, by **435.**

Corollary. *The locus of a point whose pedal triangle with regard to an equilateral triangle is described in a given direction and has a given Brocard angle, is a circle concentric with the triangle.*

498. Theorem. *In any triangle, the locus of a point whose pedal triangle has a constant Brocard angle and is described in a given direction, is a circle of the Schoute system; that is, a circle coaxal with the circumcircle and the Brocard circle.*

First, let an inversion be performed with an isodynamic point S' as center, leaving the second isodynamic point S in place. The given triangle inverts into an equilateral triangle with S as center (**492**); the circles of Apollonius invert into lines through S, and the Schoute circles into circles concentric about S. But by **204**, if a triangle and a point are subjected to an inversion, the pedal triangle of the point with regard to the triangle is inversely similar to the corresponding pedal triangle in the inverse figure. Now in the inverted figure, each Schoute circle is by **497** the locus of a point whose pedal circle has a constant Brocard angle; and since this property remains unaltered by an inversion, it is also valid in the original figure.*

GENERALIZATIONS

499. Many attempts have been made, with varying success, to generalize the Brocard geometry. The theory of similar figures which is discussed in the next chapter is evidently a generalization of certain parts, but there we find no obvious analog for the Brocard points. We summarize very briefly one or two other forms of generalization.

Theorem.† *Let P, Q be any two points, and $P_1P_2P_3$ and $Q_1Q_2Q_3$ any Miquel triangles of these points, with respect to triangle $A_1A_2A_3$.*
If P_1P meets Q_1Q at B_1, P_2P meets Q_2Q at B_2, P_3P meets Q_3Q at B_3, then B_1, B_2, B_3, P, Q lie on a circle, the generalized Brocard circle. The perpendiculars from the B-points to the corresponding base-lines meet at a point O on this circle, and parallels to the base-lines through the B-points meet at a point K on the circle. Triangles $B_1B_2B_3$ and $A_1A_2A_3$ are inversely similar. Lines through A_1, A_2, A_3, respectively perpendicular or parallel to the sides of $B_1B_2B_3$, are concurrent on the circumcircle of $A_1A_2A_3$.

Other theorems may be observed, and numerous special cases are of interest.

---

* This theorem, with an analytic proof, is due to P. H. Schoute, *Proceedings, Amsterdam Academy*, 1887–88, pp. 39–62. See also Gallatly, chap. VIII.

† J. A. Third, *Proceedings of Edinburgh Math. Society*, XXXI, 1912, pp. 17–34.

500. Another scheme, due to Lemoine, depends on a generalization of **440**.

Theorem. *Let K be any point in the plane, and let A_1K meet A_2A_3 at K', etc. Let lines through K', parallel respectively to A_1A_3 and A_1A_2, meet A_1A_2 and A_1A_3 at L_3 and M_2. Then A_1L_1, A_2L_2, A_3L_3 are concurrent at a point W, and A_1M_1, A_2M_2, A_3M_3 at another point W';* these points have many properties resembling those of the Brocard points.

501. A generalization of the Tucker circles was studied by Third.* Essentially, his circles are those circumscribing the Miquel triangles of any isogonal conjugate points P and P' (**459**). The resemblance to Tucker circles extends to many details.

502. A remarkable group of theorems by Hagge † describe a general circle in the triangle.

Theorem. *Let P be any point, P' its isogonal conjugate, and let A_1P meet the circumcircle at B_1, etc. Let the reflection of B_1 with regard to A_2A_3 be C_1, etc. Let C_1P meet the altitude A_1H at D_1, etc. Then the seven points H, C_1, C_2, C_3, D_1, D_2, D_3 lie on a circle. The point T, diametrically opposite to H, has in $A_1A_2A_3$ the position homologous to the position of P' referred to $O_1O_2O_3$.*
Again, if O_1P, O_2P, O_3P meet the nine point circle at X_1, X_2, X_3, and if Y_1, Y_2, Y_3 are the reflections of X_1, X_2, X_3 in O_2O_3, O_3O_1, O_1O_2 respectively; finally, if OO_1 and PY_1 meet in Z_1, etc., then the seven points O, Y_1, Y_2, Y_3, Z_1, Z_2, Z_3 lie on a circle.

We may note that if P is the incenter, the first theorem gives the Fuhrmann circle (**367**); if P is the median point, the second theorem gives the Brocard circle, as is seen in the following exercise; another instance is given in **477**, and others may be easily found.

* *L.c.*, XVII, 1898, pp. 70–99.
† *Zeitschrift für Math. Unterricht*, 1907–08, p. 257.

Exercise. *Prove that the vertex C, of the second Brocard triangle, is on circle $A_1O_2O_3$, and that if A_1M meets the nine-point circle at X_1, then triangles $O_2O_3C_1$ and $O_2O_3X_1$ are symmetrically congruent.*

503. Attempts to extend the Brocard geometry to the quadrangle are only moderately successful. In order to have results of any value, we must restrict ourselves to the so-called harmonic quadrangle, whose characteristic property is that it is inscribed in a circle, and the products of opposite sides are equal. We have seen (**133**) that the vertices of a harmonic quadrangle are inverse to those of a square; again (**200**) that if $A'B'C'D'$ is a square, P any point, PA', PB', PC', PD' meet the circle $A'B'C'D'$ in vertices of a harmonic quadrangle.

504. Theorem. *In a harmonic quadrangle $ABCD$, there exist a point P, and a point Q, such that*

$$\angle PAB = \angle PBC = \angle PCD = \angle PDA$$

$$\angle QBA = \angle QCB = \angle QDC = \angle QAD$$

*The four points of intersection of such lines as AP and BQ lie on a circle through the center O of the circle through A, B, C, D. If OK is a diameter of this circle, K is the intersection of the diagonals, and corresponds to the symmedian point. Lines through K and parallel to the sides meet the other sides in eight points on a circle whose center is the midpoint of OK. Analogues of the other Tucker circles may be worked out. K is that point for which the sum of the squares of the distances from the sides of the quadrangle is a minimum.**

Exercise. The reader is invited to furnish proofs of the unproved propositions in the following sections: **480, 481, 482, 484, 485, 489, 490, 492–496, 499–504.**

* Cf. Tucker, *Educational Times*, Reprint, 44, p. 125; Neuberg, *Mathesis*, 1885, p. 202; Gob, *Cong. de Marseille*, 1891; Eckhardt, *Archiv der Math. und Physik*, 13, p. 12; *Zeitschrift für Math und Phys. Unterricht*, 36, 1905, p. 409.

CHAPTER XVIII

THREE SIMILAR FIGURES

505. When three similar figures are constructed on the three sides of a triangle as bases, there is an intimate relation between these figures and the Brocard configuration. We shall study this problem in some detail, investigating the location of corresponding points that are collinear, or of the points at which corresponding lines may be concurrent, and determining the centers of similitude. We then take up the more general problem, of three directly similar figures lying in any positions in a plane. We find that the centers of similitude and the other notable points suggest a group of theorems, of which the Brocardian theorems are a special case. It seems best to treat the special and the general cases separately by different methods; the parallelism between the two will be manifest at all times.

506. Let three similar figures be constructed, so that the lines A_2A_3, A_3A_1, A_1A_2 are homologous members. A familiar example is the Euclidean proof of the theorem of Pythagoras, wherein squares are drawn externally on the sides of a right triangle. Other sets of homologous points with which we are familiar are the mid-points of the sides, and the vertices of the first Brocard triangle (**461**).

Theorem. *The centers of similitude of the three figures, two by two, are the vertices of the second Brocard triangle C_1, C_2, C_3. The ratio of similitude with respect to C_1 is a_3/a_2, the angle of similitude* $180 - \alpha_1$.

For we saw in **463** that triangles $C_1A_1A_2$, $C_1A_3A_1$ are similar, so that C_1 is self-homologous in the similar figures on A_1A_2 and A_3A_1. The circles used in the proof of the theorem of **32** are here the c-circles c_1, c_1'; whence we have a direct proof resting on that fundamental theorem.

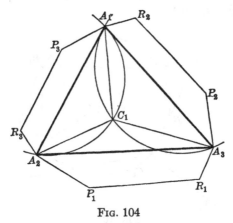

Fɪɢ. 104

507. Theorem. *The median point of the triangle of any homologous points is at M (**358**); if any three homologous points P_1, P_2, P_3 are collinear, their line passes through M, and $\overline{MP_1} + \overline{MP_2} + \overline{MP_3} = 0$.*

508. We next consider the possibility of three homologous lines being concurrent. For instance, we are already acquainted with the perpendicular bisectors of the sides, and with the lines from the vertices to either Brocard point.

Theorem. *If three homologous lines are concurrent, their point of concurrence is on the Brocard circle, and they pass respectively through the vertices of the first Brocard circle.*

Let us first consider three homologous lines parallel to the sides; then their distances from the latter are proportional to a_1, a_2, a_3; and if three such lines meet, their common point must be the symmedian point K (**342**), and they are the lines KB_1, KB_2, KB_3.

Three homologous lines that meet the sides, say at P_1, P_2, P_3 respectively, make equal angles with them. If three such lines are concurrent, say at a point U, then A_1, P_2, P_3, U are on a circle; and this circle passes through the center of similitude C_1 (**33**). Hence

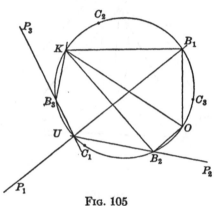

FIG. 105

$$\angle P_3UC_1 = \angle P_3A_1C_1;\quad \text{similarly,}\quad \angle C_2UP_3 = \angle C_2A_2P_3$$

Adding, and remembering that C_1, C_2, C_3 are on the respective symmedians,

$$\angle C_2UC_1 = \angle C_2A_2, C_1A_1 = \angle KC_2, KC_1 = \angle C_2KC_1$$

and we see that C_1, C_2, K, U are concyclic; but the first three are on the Brocard circle, and therefore the first part of the theorem is proved. Let UP_1 meet this circle again at Q_1, then

$$\angle UQ_1K = \angle UC_2K = \angle UC_2A_2 = \angle UP_3A_2 = \angle UP_1A_3$$

so that Q_1K is parallel to A_2A_3, and Q_1 coincides with B_1.

509. Theorem. *Conversely, the lines connecting any point of the Brocard circle with the vertices of the first Brocard triangle are homologous.*

For if B_1U, B_2U, B_3U, through a point U of the Brocard circle, meet the sides at P_1, P_2, P_3, we see easily that they make equal angles with the sides, and that therefore triangles $B_1O_1P_1$, $B_2O_2P_2$, $B_3O_3P_3$ are similar.

Corollary. *Three homologous lines through the points B_1, B_2, B_3, meet on the Brocard circle.*

510. Theorem. *The pedal triangle of any point on the Brocard circle has the Brocard angle ω.* (Cf. **498.**)

For in the above discussion, $P_1P_2P_3$ is a Miquel triangle of U; hence the result by **476.**

We consider next three homologous lines in general.

511. Theorem. *If a triangle is formed by three homologous lines, its symmedians pass through the vertices of the second Brocard triangle and meet at a point on the Brocard circle; the center of similitude of the triangle with the given triangle is also on the Brocard circle.*

For through homologous points P_1, P_2, P_3 on A_2A_3, A_3A_1, and A_1A_2 respectively, let three homologous lines l_1, l_2, l_3 be drawn, meeting at L_1, L_2, L_3. (We recognize that these points are not generally homologous.) Obviously A_1, P_2, P_3, L_1, C_1 are concyclic;

$$\measuredangle\ P_3A_1C_1 = \measuredangle\ P_3L_1C_1, \text{ that is, } \measuredangle\ A_2A_1K = \measuredangle\ L_2L_1C_1$$

But triangles $A_1A_2A_3$ and $L_1L_2L_3$ are similar; if U denotes the symmedian point of the latter,

$$\measuredangle\ A_2A_1K = \measuredangle\ L_2L_1U$$

Hence L_1, C_1, U are collinear, and C_1 is on the symmedian L_1U. Since $L_1L_2L_3$ is similar to $A_1A_2A_3$,

$$\measuredangle\ L_1UL_2 = \measuredangle\ A_1KA_2, \quad \measuredangle\ C_1UC_2 = \measuredangle\ C_1KC_2,$$

and U is on the circle C_1C_2K. Finally, the center of similitude lies on a circle through homologous points U, K and the point of intersection C_1 of homologous lines A_1KC_1, L_1UC_1; this is the Brocard circle.

Corollary. *The center of similitude of any two triangles whose sides are homologous lies on the Brocard circle.*

512. We next investigate triads of homologous points that are collinear. We have already seen that such points are collinear with the median point.

Theorem. *Three homologous collinear points lie respectively on the circles through M and two of the vertices of the second Brocard triangle (the circles of McCay, cf. 485).*

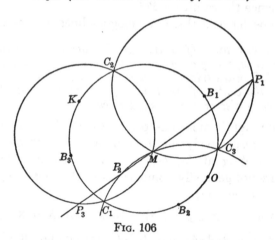

Fig. 106

The point C_1 is a center of similitude, therefore if P_1, P_2, P_3 are homologous, triangle $C_1P_2P_3$ is similar to the fixed triangle $C_1B_2B_3$. Hence if $P_1P_2P_3$ passes through M (**507**),

$$\angle\, C_1P_2M = \angle\, C_1P_2P_3 = \angle\, C_1B_2B_3 = \angle\, C_1C_3B_3 = \angle\, C_1C_3M$$

(**465**) so that P_2, C_1, C_3, M are concyclic.

513. Theorem. *Conversely every line through M cuts the circles of McCay in homologous points.*
If a point moves on a McCay circle, the homologous points trace the other McCay circles so as to be collinear with M, and $\overline{MP_1} + \overline{MP_2} + \overline{MP_3} = 0$. On a tangent to a McCay circle

at M, the other circles cut equal chords. If MC_1 cuts the McCay circle MC_2C_3 at C_1', then $\overline{MC_1'} = 2\overline{C_1M}$; C_1' is the point which, referred to the figure on A_2A_3, corresponds to the self-homologous point C_1 of the other two figures. Triangle $C_1'C_2'C_3'$ is homothetic to $C_1C_2C_3$.

514. Theorem. *If the lines connecting the vertices $A_1A_2A_3$ to three homologous points $P_1P_2P_3$ are concurrent, then either P_1, P_2, P_3 are on the perpendicular bisectors of the sides (357) or else they are on the respective Neuberg circles (480) and the connectors are parallel.*

Unfortunately the proof of this excellent theorem is long and involved;* as it is in no sense instructive, we omit it.

The converse theorem leads to the following properties of the McCay circles.

515. Theorem. *If parallel lines are drawn through the vertices they meet the respective Neuberg circles in homologous points.*

Theorem. *A line through M, parallel to the connectors A_1P_1, A_2P_2, A_3P_3, passes through the median points of the homologous triangles $A_2A_3P_1$, $A_3A_1P_2$, $A_1A_2P_3$. That is, the loci of these median points are the respective circles of McCay.*

For a line through M, parallel to A_1P_1, trisects P_1O_1, which is a median of triangle $A_2A_3P_1$.

Corollary. *The mid-point of the side A_2A_3 is the external center of similitude for the opposite Neuberg circle and McCay circle, the ratio being $3/1$. The center Y_1 of the McCay circle MC_2C_3 is on OO_1,*

$$\overline{O_1Y_1} = \frac{a_1}{6}\cot\omega$$

and the radius is $r = \frac{a_1}{6}\sqrt{\cot^2\omega - 3}$.

The radius of the McCay circle is the mean proportional between $\overline{Y_1O_1}$ and $\overline{Y_1B_1}$. The vertices of the D-triangle (477) lie on the respective McCay circles.

* See Emmerich, *l.c.*, §167.

516. We now turn to the more general problem of the mutual relationships of three directly similar figures in a plane. We have seen in Chapter II that two similar figures determine a unique center of similitude, or self-homologous point; and in the first part of this chapter we have discussed three similar figures having three sides of a triangle as homologous bases. We find that many of the theorems apply with some modification to the more general case.*

Let us first recall the method of determining the self-corresponding point of two figures; if two homologous segments MN, $M'N'$ are chosen as base lines, and these lines meet at X, then the second intersection C of circles $MM'X$ and $NN'X$ is the center of similitude. Moreover, if any lines MP, $M'P$ meet at P on the circle $MM'C$, they are homologous; if CH and CH' are the perpendiculars from C to these lines, the figure $CHPH'$ is of fixed form.

Considering now the case of three figures, we may think of them as determined by a set of homologous segments or baselines, M_1N_1, M_2N_2, M_3N_3. To avoid troublesome limiting cases, we shall assume that none of these are parallel, that no two of the six points coincide, and that the three centers of similitude are not collinear. We may also require that the lines M_1N_1, M_2N_2, M_3N_3 are not concurrent, but form a triangle $L_1L_2L_3$. We denote the center of similitude of II and III by C_1, or sometimes, to emphasize that it is self-homologous in these figures, by C_{23}; similarly the other centers of similitude are C_2 and C_3, or C_{31} and C_{12}. The circle $C_1C_2C_3$ is called the circle of similitude. In triangle $L_1L_2L_3$, denote the Miquel point of $M_1M_2M_3$ by M, that of $N_1N_2N_3$ by N. Then, for instance, L_1, M_2, M_3, C_1, M lie on one circle, and L_1, N_2, N_3, C_1, N on another circle.

517. Theorem. *The circle of similitude passes through the*

---

* Cf. Simon, *l.c.*, pp. 171–172.

points M and N.   The lines L_1C_1, L_2C_2, L_3C_3 are concurrent on the circle of similitude.

For let L_2C_2 meet L_3C_3 at U, then

$$\angle C_2MC_3 = \angle C_2M, L_2C_2 + \angle L_2C_2, L_3C_3 + \angle L_3C_3, MC_3$$

$$= \angle MC_2L_2 + \angle L_3C_3M + \angle C_2UC_3$$

$$= \angle MM_1L_2 + \angle L_3M_1M + \angle C_2UC_3$$

$$= \angle C_2UC_3$$

Similarly, $\angle C_2NC_3 = \angle C_2UC_3$

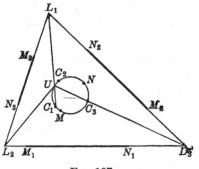

FIG. 107

Hence the circle through C_2, C_3, and M passes also through U and N, and therefore through C_1 as well.

518. Theorem. *The triangle of any three homologous lines is in perspective with the triangle of similitude, and the locus of the center of perspective is the circle of similitude.   The distances from the center of perspective U to the homologous lines are proportional to the scales of the figures.   (Cf. 511.)*

If lines UL_1, UL_2, UL_3 are divided proportionally at L_1', L_2', L_3', then $L_2'L_3'$, $L_3'L_1'$, $L_1'L_2'$ are parallel to L_2L_3, L_3L_1, L_1L_2 respectively, and homologous.   In particular, lines through U, parallel to the respective base-lines, are homologous.   Conversely, if three homologous lines are concurrent, their common point lies on the circle of similitude.

519. Theorem. *Three concurrent homologous lines pass respectively through three fixed points on the circle of similitude, called the invariable points.*

Let XP_2, XP_3, homologous lines of II and III, whose intersection X is on the circle of similitude, meet that circle again at T_2 and T_3. Then arcs C_1T_2 and C_1T_3 are constant in magnitude and direction, and T_2 and T_3 are fixed points. Similarly X_1P_1 passes through a fixed point T_1.

Corollary. *The invariable points are homologous, and their triangle is inversely similar to any triangle whose sides are homologous lines (461). Any set of lines joining the invariable points to a point of the circle of similitude are homologous.*

520. *The triangle of similitude and the invariable triangle are in perspective at a point Q (465).*

For let us write C_{23} instead of C_1, to remind ourselves that it is self-homologous in II and III; and let the homologous point in I be C_1'. Let T_2C_2 meet T_3C_3 at Q, then C_1' lies on the circle C_2C_3Q. For

$$\angle\, C_2C_1'C_3 = \angle\, C_2C_1'T_1 + \angle\, T_1C_1'C_3$$
$$= \angle\, C_2C_{23}T_3 + \angle\, T_2C_{23}C_3$$
$$= \angle\, C_2T_2T_3 + \angle\, T_2T_3C_3$$
$$= \angle\, C_2T_2,\, T_3C_3 = \angle\, C_2QC_3$$

Also, C_1' and Q lie on the line $C_{23}T_1$. For

$$\angle\, C_3T_1C_1' = \angle\, C_3T_3C_{23} = \angle\, C_3T_1C_{23}$$
$$\angle\, C_3C_1'Q = \angle\, C_3C_2Q = \angle\, C_3C_2T_2 = \angle\, C_3C_{23}T_2 = \angle\, C_3C_1'T_1$$

The intersection of the line C_1QT_1 with the circle QC_2C_3 is the point C_1' which corresponds to C_{23} (513).

521. Theorem. *Given two triads of non-concurrent homologous lines, forming similar triangles; their respective centers of perspective are homologous with respect to these triangles,* *and their center of similitude lies on the circle of similitude* (**512**).

522. Theorem. *If P_1, P_2, P_3 are homologous points in the similar figures, the circles $P_1C_2C_3$, $P_2C_3C_1$, $P_3C_1C_2$ are concurrent at a point P.*

For we saw in **516** (end) that circles $P_2P_3C_1$, $P_3P_1C_2$, $P_1P_2C_3$ are concurrent; hence the result, by **188** *e*.

Theorem.

$$\measuredangle\, C_2QC_3 = \measuredangle\, P_2P_1P_3 + \measuredangle\, C_2P_1C_3$$

523. Theorem. *If homologous points move in such fashion that $\measuredangle\, P_2P_1P_3$ is constantly equal to a given angle α, each lies on a fixed circle; the locus for P_1 is that circle through C_2 and C_3 for which*

$$\measuredangle\, C_2P_1C_3 + \alpha = \measuredangle\, C_2QC_3$$

while P_2 and P_3 lie on the homologous circles. In particular, the loci of collinear homologous points are the circles C_2C_3Q, C_3C_1Q, C_1C_2Q.

Theorem. *Any line through three homologous points passes also through Q.*

For $\quad \measuredangle\, C_3QP_1 = \measuredangle\, C_3C_1'P_1 = \measuredangle\, C_3C_1P_2 = \measuredangle\, C_3QP_2$

524. Theorem. *There is a single triad of homologous points which are the vertices of a triangle similar to a given triangle.*

Let it be required to find a triangle of homologous points $P_1P_2P_3$, similar to a given triangle $V_1V_2V_3$. The locus of each vertex, when one angle, say $P_2P_1P_3$, equals the given angle $V_2V_1V_3$, is a circle. If now a second angle is assigned, we have a second triad of locus circles; and it is easily seen that these meet in one set of homologous points. A different and somewhat longer proof is needed to establish the analogous theorem in the special case that V_1, V_2, V_3 are collinear:

525. Theorem. *There is a single triad of homologous points whose distances from one another are as the distances of any three given collinear points.*

526. Exercises. *a. Find the locus described by each of three homologous points in three similar figures, if the ratio of $\overline{P_1P_2}$ to $\overline{P_1P_3}$ is constant; if the area of triangle $P_1P_2P_3$ is constant; if the length of $\overline{P_2P_3}$ is constant.*

b. Determine how far the fundamental elements of a system of three similar figures may be taken arbitrarily. For instance, show that if two triangles are in perspective and are inscribed in the same circle, either may be taken as the triangle of similitude, and the other as the invariable triangle, of such a system.

c. Establish completely the parallelism between the general theory here discussed and the special theory in the first part of the chapter. The first Brocard triangle is the invariable triangle, the second the triangle of similitude, the circles of McCay the circles QC_2C_3, etc.; and so on.

d. In the general case, the condition for the existence of a triangle $A_1A_2A_3$ on which the similar figures are based, is that the point Q be the median point of the invariable triangle.

Exercise. In this chapter, proofs are to be furnished by the reader in the following sections: **507**, **509**, **513**, **(514)**, **(515)**, **518**, **519**, **521-523**, **(525)**, **(526)**.

INDEX OF NOTATION FOR
THE TRIANGLE

THE following symbols are used consistently throughout the book. Unfortunately it is necessary to use some symbols in different parts of the work with different meanings; the following includes all standardized symbols.

It is understood that generally the letter designating any point, when written with a subscript, denotes the foot of the perpendicular from that point on the designated side of the triangle. (Thus K_1, K_2, K_3 are the vertices of the pedal triangle of the point K.) But this is not an invariable convention; the subscripts are frequently used in other senses (as in the case of N, N_1, N_2, N_3).

SYMBOL	MEANING	PAGE REFERENCE
A_1, A_2, A_3	vertices of given triangle	8
a_1, a_2, a_3	lengths of sides	8
a_1, a_2, a_3	its angles	8
B_1, B_2, B_3	vertices of first Brocard triangle	278
C_1, C_2, C_3	vertices of second Brocard triangle	279
D_1, D_2, D_3	vertices of an associated triangle	284
F	mid-point of OH	195
H	orthocenter	9
H_1, H_2, H_3	feet of the altitudes	9
h_1, h_2, h_3	lengths of the altitudes	9
I	incenter	9
J', J'', J'''	excenters	182
K	symmedian point	213
L_1, L_2, L_3	centers of circles of Apollonius	294
M	median point	9
m_1, m_2, m_3	lengths of medians	9

Symbol	Meaning	Page Reference
N	Nagel point	225
N_1, N_2, N_3	centers of circles of Neuberg	287
O	circumcenter	8
O_1, O_2, O_3	mid-points of sides	8
R	radius of circumcircle	8
R, R'	isogonic centers	218
S, S'	isodynamic points	295
S	Steiner point	281
s	half perimeter	9
T	Tarry point	282
Z	mid-point of OK	273
Δ	area of given triangle	9
ρ	radius of incircle	9
ρ_1, ρ_2, ρ_3	radii of excircles	182
Ω, Ω'	Brocard points	264
ω	Brocard angle	264

INDEX